GOING ROUND THE BEND
# IN SEARCH OF
# Old Tractors

GOING ROUND THE BEND
# IN SEARCH OF
# Old Tractors

Ian M. Johnston

NEW
HOLLAND

First published in 2014 by New Holland Publishers Pty Ltd
London • Sydney • Auckland

The Chandlery Unit 114 50 Westminster Bridge Road London SE1 7QY United Kingdom
1/66 Gibbes Street Chatswood NSW 2067 Australia
218 Lake Road Northcote Auckland New Zealand

www.newhollandpublishers.com

A record of this book is held at the British Library and the National Library of Australia.

ISBN 9781742576084

Managing Director: Fiona Schultz
Publisher: Diane Ward
Designer: Andrew Quinlan
Photographs: Ian M. Johnston, Margery Daw
Production Director: Olga Dementiev
Printer: Toppan Leefung Printing Ltd (China)

10 9 8 7 6 5 4 3 2 1

Keep up with New Holland Publishers on Facebook

www.facebook.com/NewHollandPublishers

I dedicate this book to my adorable wife Margery Daw.  For over half a century she has patiently accompanied me on my research treks around the world, encouraging me when I became despondent, insisting I pause for meal breaks, checking the camera is ready to go, writing down notes on my behalf and providing inspiration when I ran out of ideas.

Together, we have visited no less than thirty four countries, the majority of which we traversed in a rental car and simply proceeded to find our own way around.  Organised group tours are simply not appropriate for carrying out investigative research into the history of classic tractors.

Thank you Margery.  Without your cherished support there would be no *GOING ROUND THE BEND IN SEARCH OF OLD TRACTORS.*

# AUTHOR'S NOTES.

Whilst I have endeavoured to achieve total accuracy of data and information, the passage of time can dim the memory. I therefore request a degree of tolerance should I have ever strayed inadvertently from the path of exactitude.

On several occasions I have made reference to tractors and farming in Scotland. Please excuse and accept this as being expressions of nostalgia by an ageing Scotsman.

After wrestling with the concept, I decided to utilise either metric or imperial performance and dimensional measurements, whichever I thought appropriate to the related text. It simply did not feel right to convert the early standard of 'number of horses replaced' to kilowatts and kilojoules.

I apologise for the indifferent quality of some of the graphics, most of which were taken prior to the introduction of the digital camera. Unless otherwise stated, the photos were taken by myself, using a variety of cameras.

I am grateful to the management of Greenmount Press for their consent to my request to utilize in GOING ROUND THE BEND IN SEARCH OF OLD TRACTORS certain extracts from material previously published in *The Australian Cotton Grower; Australian Grain;* and *Australian Sugarcane.*

I am indebted to New Holland Publishers who have again exhibited confidence in my work. I thank them for this, for an author without a publisher's backing and support – is but a dreamer!

My thanks also go out to my old and new friends who have contributed so generously with their knowledge and suggestions, making this book possible. I also wish to extend my gratitude to the thousands of readers around the world who have supported my writing in the past. It is this sort of encouragement that inspires me to continue pushing ahead with my researching and writing!

Ian M. Johnston.

# CONTENTS.

# PREFACE.

NOT ALL BAD NEWS.

Recently at a tractor rally, I was amused to overhear a bespectacled academic looking young father tell his ten year old son that in days long ago farmers actually *worked* these ancient tractors. The boy shook his head in wonderment at the very thought of such a thing.

It made me think too. Because to me the tractors seemed perfectly familiar and normal. But of course I grew up with these fine old machines. So does that render me ancient, like the tractors? Certainly not!

However I do have to agree that I am not *quite* as young as I used to be. (A really profound statement!) To be honest, it is not all bad news. I am offered Senior Citizen's discounts in shops, and am often addressed by respectful assistants as 'Sir', but secretly would prefer 'Mate'. Road recovery service 'chaps' (a good old fashioned term) will happily change a tyre, as the physical activity involved would obviously be beyond the ability of such an 'aged' person as myself. How good is *that*?

It is noticeable that I no longer leap with alacrity up into the tractor seat, but rather tend to haul myself onboard, then pause a moment or two in order to recover my breath. Following which, I no longer roar down the paddock, but move off sedately in the manner of a well disciplined chauffeur putting the Rolls into motion.

I could also mention such tedious things as thickening of the joints, thickening of the midriff and indeed possibly the thickening of my little grey cells. But I won't, for to do so would amount to a confession that I am 'getting on', as my grandchildren so euphemistically, but somewhat uncharitably, put it.

So far I have not been approached by boy scouts with offers to assist me across the road. That is definitely encouraging. And when at the wheel of our ostentatious bright red roadster, hoons in twenty year old Mazdas pull alongside keen for a burn. Obviously *they* don't think I am past it. Could they possibly think I am cool? Gee I hope so.

THE GOOD NEWS.

Despite all this, Margery and I enjoy a fairly frantic social life with our tractor and classic car compatriots. Also, we continue to travel extensively, including overseas where we rent a car and set off into the sunset in pursuit of rare tractors, often in remote places.

The important thing is, we continue to shock our grandchildren with our ungrandparent like goings on -

guaranteed to be the best way to remain young at heart!

So what is the point I am making?

When in the company of classic tractor enthusiasts, most of whom will be younger than me (usually a *lot* younger), should a disputation arise concerning an historical or technical fact relating to a tractor of bygone days, my younger associates will normally turn to me for arbitration. This of course does wonders for my not inconsiderable egotistical proclivity!

They reckon that someone as prehistoric as me must have been around at the same time as the tractor in question and therefore will likely know all about it. And, they are right! Well – usually.

Think about it. Even as a young tousle haired schoolboy in Scotland during the mid 1940s, I was driving tractors, and not just playing around – but actually ploughing and things. Well there was a war on and country boys had to muck in and lend a hand. There was an acute shortage of young farm workers, many of whom had been dragged off to the recruiting centre, issued with a scratchy uniform, plus a 303 rifle and told to go and defend The Empire!

Then there were the old men, who were too old to go to war. Many had served in The Great War (not sure what was *great* about it) and were keen to resume where they had left off doing battle with the Hun. But the recruiting officers rather indelicately told them to go back to their rocking chairs and leave the fighting to *young* men.

As a consequence, having no desire to sit out the war in a rocking chair, many pensioned-off farm labourers tended to turn up at a farm to see if they could lend a hand. Their offered services proved a considerable boon to the farmers, who were all under pressure from the wartime Ministry of Food to produce more and more crops to feed British citizens, barely surviving from the results of strict food rationing.

These elderly farm workers were frequently bent and gnarled from the results of years of gruelling hard labour, stemming back to when they were mere children. Despite this, they brought with them skills and crafts including stack

*An early harvest scene of a binder being hauled by a three horse team, whilst alongside a team of labourers pick up the sheafs and form stooks, enabling the grain to be dried prior to threshing.*

building, thatching, a love and compassionate understanding of farm animals, and in many instances the ability to plough a straight furrow with a two horse mouldboard plough.

There were also those who had a knowledge of tractors in the early nineteen hundreds, at a time when farm tractors first disturbed the tranquility of the countryside as they clattered across fields. Some could even boast of having mastered the daunting task of driving these cantankerous machines.

All old men like to yarn on about 'the good old days'. In modern times, younger folk seem not to have the time to listen, which is their misfortune. But on a farm in Scotland back in the 1940s, there was any amount of time to listen!

I can recall spending hours listening enthralled to first hand accounts of how the introduction of the first of the internal combustion engined tractors, such as the International Titans, Marshall Colonials, Ivels, Rumley Oil Pulls, Saundersons, etc., slowly but surely began the demise of the steam traction engine and brought a challenge to the heavy draught horse. I heard of the fierce opposition to the arrival of 'these new dangerous and accursed monsters'. Many were the tales of horses bolting at the sight of 'such hideous contraptions'.

Significantly and indeed fortuitously (as it turns out), I was acquiring living history about farm tractors from the mouths of those who were there!

SO WHAT HAPPENED NEXT?

The next bit is less exciting stuff, but bear with me as it sort of puts things into perspective.

In 1952, at age seventeen, I arrived in Australia from Scotland for a two year stint working on varying types of rural properties. In order to be accepted into The East of Scotland Agricultural College one was obliged to complete a two year practical course in agriculture, which could be served abroad. The problem was (and is) I fell in love with Australia and convinced my parents that this country offered better opportunities than Scotland.  So I remained!

My youthful interest in tractors has abided with me throughout my life.  I have earned a living firstly by driving tractors, followed by an involvement with the  marketing of tractors, and more recently - farming with tractors.

Today my 'work' revolves around writing about tractors. Indeed I have been scribbling magazine tractor articles for more years than anyone can remember. In fact one editor, whose publishing house is located in Toowoomba, Queensland, following an association extending around fifteen years, now regards me in the same way he would his most comfortable pair of elastic sided Baxters or his favourite chair in front of the telly.  In other words I seem to be just a familiar part of the scene and therefore happily tolerated!

On average, about once every three years, I drive up to

the office at Toowoomba and throw my hat in the door. Reassuringly, the folk there always remember who I am and in no time a cup of coffee is forthcoming. If it is lunchtime I am treated to a meal at a nearby sporting club. The conversation seems to always drift in the direction of rugby and/or fishing. Being a polite sort of cove I listen intently and even manage some comments that are not too inane. You see I am a League follower and a non-fish person, unless the things are grilled and served up with chips.

The writing, which I thoroughly enjoy, actually works out well for me. Thanks to the marvels of the internet, all the 'work' is done in my farm office. I can gaze out at the trees, paddocks and things for inspiration. Or play truant and go and tinker with a tractor or an old car, or go and count my vast herd of four Belted Galloways. Or maybe even just watch the grass grow!

In addition to my magazine efforts, I have had seven books published and sold worldwide. This is an indication of the interest in old tractors, rather than the calibre of my work.

The point I wish to make is – the catalyst that enabled me to become a writer of tractor history, is the experience I had with these wonderful old characters in Scotland all these years ago. Although not realising it at the time, I was truly fortunate and owe a great deal to them, as I do to the equally fascinating individuals with whom I worked in the 1950s in Australia, especially during my period as a roving tractor driver and a shearer's rouseabout.

Without the 'blessings' that accompany maturity, one cannot have the long tunnel of memories and the gifts and consolations this brings.

So you see, quite clearly, sharing a commonality with the old tractors has its advantages and is not all bad news!

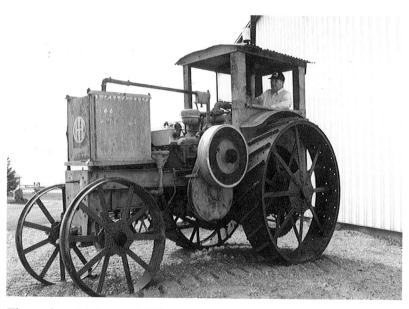

*The author driving an 1917 International Titan 15-30 which featured a double chain drive to rear wheels. One of a range of early American tractors featured at The Saskatchewan Western Development Museum, Saskatoon, Canada. (Photo M. Daw)*

# 'HP' - WHAT DOES IT MEAN?

Recently I asked a diverse group of individuals to tell me the first thought that comes to mind when they hear the initials 'HP'. A rather frivolous request you may think. But have faith in this scribe – there was an intellectual purpose behind this somewhat curious request. You will see how these wonderful two letters of our alphabet can conjure up entirely differing interpretations.

For example, predictably the majority of the group immediately rattled off 'HP Sauce' which, as everyone knows, for nigh on a century has been embellishing juicy roast beef and succulent T-bone steaks. It has even competed with tomato sauce for the enhancement of the plebeian sausage!

On the other hand, when confronted with HP, the cerebral minds of young fledging bespectacled technocrats automatically switch to Hewlett Packard and that company's range of computers, printers and all the rest of its doodahs.

HP of course meant hire purchase to the post war families of the 1950s and 60s, for whom the possession of an Admiral TV,

Silent Knight fridge or a Victa lawn mower inevitably involved a visitation to H. G. Palmers, Grace Brothers or their equivalents, followed by the signing of the dreaded hire purchase documents.

Tractor historians (worth their salt) would immediately conjure up images of Hart Parr tractors. To a steam boffin, HP would likely be construed as high pressure, but to a bowler hatted UK MP, HP (couldn't resist that) means only one thing – Houses of Parliament. Which brings us right back to HP Sauce!

However, there is an elite group of humanity to whom the digits HP acquire a very different interpretation to those mentioned thus far. I am of course referring to our cherished farmers and that all embracing terminology *HORSE POWER!* Yes, even in this multifarious era of metrication, thanks to the Yanks and their tractor power ratings, good old horse power remains as relevant today to farmers as it did a century ago. Well, not quite!

To explain and extrapolate it is first necessary to examine the origins of horse power.

## JAMES WATT.

The dawning of the Twentieth Century saw the commencement of the tranquility of the countryside being disturbed by the raucous sounds of the first tractors. The few farmers who could afford and were bold enough to contemplate the purchase of one of these giant clattering machines, were naturally concerned with the tractor's ability to pull a plough in comparison with their proven and trusted horse team.

In response, tractor manufacturers arrived at a simple comparison criteria. They stated that a particular model of tractor had the ability to replace a certain number of horses. The terminology utilised by most tractor companies was *'the number of horses replaced'*. Farmers felt comfortable with, and could relate to, this type of down-to-earth analogy.

However as tractor engine and transmission technology advanced, the more scientific and exacting *Horse Power* ratings were universally applied to tractors. Which begs the question – what is a horse power? (You'll wish you had never asked!)

The Scottish inventor James Watt (1736-1819), of steam engine fame, experienced frustrating opposition to his doctrine when he stated that in coal mines steam engines were more practical than horses, whilst engaged in the heavy task of raising coal to the surface.

In a practical test he demonstrated that a horse, walking at the rate of 2½ m.p.h. could draw a mere 150 pounds of coal (by means of a rope fed through a pulley) vertically at the rate of 220 feet per minute. This equated, so he determined, to 33,000 pounds, raised vertically to the height of 1 foot in one minute as being one horsepower!

Quite honestly, my limited expertise in the field of mathematics renders me incapable of establishing how my fellow Scot arrived at this conclusion. (Had Watt been enjoying a wee dram or two whilst making his calculations?) Indeed many mathematicians of repute are similarly perplexed. But be that as it may, Watt's horsepower rating has been and is the basis for a variety of enduring power related measurements, which include an entire range of horsepower standards.

*James Watt*

Boffins today are generally in agreement that 1 horsepower equates 550 foot-pounds per second or 745.56 watts, or one joule per second!  (Gosh).

So what has all the foregoing mumbo-jumbo got to do with farming and in particular early tractors?  Frankly – I haven't a clue!  So let us escape from these realms of theory into the *real* world of tractors and their relevant horse powers.

And to forestall an avalanche of protesting letters, let me emphasise that I am aware that we should all be figuring in metric – kilowatts and so forth, but this is an epistle dealing with *horse powers* and farmers in the majority still refer to the *horsepower* of their tractors.

## HORSEPOWER TYPES.

<u>Indicated horsepower</u> is of little interest to farmers, as it is purely a mathematical figure which indicates the power theoretically developed within the cylinder(s) of an engine.

<u>Brake horsepower</u> is the power developed by an engine at a specific r.p.m. available at the fly wheel and is measured by the resistance to a brake.

<u>Belt horsepower</u> is the actual horsepower developed at the belt pulley and is generally less than the brake horsepower owing to the frictional losses through the gears.  However, in some early tractors where the pulley is a fixture on the crankshaft, no frictional losses occur. Although not relevant

today, belt horsepower rating was important when most tractors were routinely used to drive ancillary machinery such as threshers, pumps, etc. via an endless belt attached to the belt pulley.

Drawbar horsepower was used extensively up until the 1960s and indicated the available horsepower actually at the rear hitch point, but did not allow for wheel slip caused by tyre conditions or soil variances.

Power take off (PTO) horsepower has replaced the belt horsepower in modern times. PTO shafts now do the job of the endless belt and are used mainly for supplying energy to

*This 1912 International Mogul is part of a collection held in a museum at Capella, Queensland. A farmer would have been told that this tractor was capable of replacing a twelve horse team. The author had the opportunity of driving this tractor whilst it was pulling a six disc Sundercut plough. (Photo M. Daw)*

front and rear mounted implements. This measurement is of the power actually available at the shaft and is the figure most relevant to the rating of modern tractors.

RAC horsepower is of no relevance to tractors as it was a figure determined by The Royal Automobile Club and used for taxing British motorists. For the record, the formulae was as follows: number of cylinders squared x number of cylinders x 0.4.

Added to the above is the need to know if the horsepower in question is *maximum* or *rated*. Maximum is determined as being the peak figure achieved for only a brief duration. Rated is a figure that can be sustained over a continuing period.

## DRAWBAR PULL.

Up until the 1970s, *drawbar pull* was an excellent and widely used measurement to compare a tractor's field performance alongside those of different models or makes. It was measured in pounds pull at a given speed. Quite often the results indicated that horse power was not the determining factor in a tractor's ability to efficiently pull a specific soil engaging implement such as a plough. Rather, drawbar pull illustrated the ability of the tractor to transfer its engine horsepower to the ground, taking into consideration, weight, traction (wheel spin), gear ratios and engine torque.

Thus a 70 horsepower tractor might well have been capable

of out-pulling a different make of 70 horsepower tractor, the drawbar pull being the determining factor.

During the 1970s, U.S. tractor manufacturers seemed to become obsessed with their desire to extract the maximum brake horsepower from their engines, very often at the expense of overtaxed transmissions. Cumbersome counterweights were attached in an endeavour to reduce wheel spin.

The European manufacturers adopted a different and more rational approach. They simply produced more solidly built tractors with better balanced engines having the emphasis on torque rather than horsepower.

*A fine example of a 1914 Minneapolis 20-40 which featured a 4 cyl. engine and was rated as 20-40 h.p. The 20 represented drawbar h.p. and the 40 represented engine h.p. Photographed at The Reynolds-Alberta Museum, Wetaskiwin, Canada.*

Wheel spin of up to 5% has been proved to be the optimum for pneumatic tyred tractors. Indeed 5% slippage is not discernible to the eye. It follows therefore that if wheel spin can be observed by simply looking, then it exceeds the 5% figure and the tractor is not operating at its maximum efficiency and will be applying excessive stress to the tyres and transmission.

## CONCLUSION.

Back in the '60s, canny farmers who may have been dithering whether to purchase a Massey Ferguson 65, a Fiat 513R or a Fordson Super Major, all similarly powered tractors, would have tossed the coloured glossy brochures aside and insisted upon a comparative demonstration. They would have walked alongside a ploughing tractor to see if they could observe wheel spin. They would also have understood the significance of drawbar pull and its relation to horsepower.

Certainly other factors had to also be considered. How did the price compare? What sort of a trade-in price would be offered on the old International W30? Was the local dealer a good bloke? Did he support your footy team? Would he deliver the tractor with a full tank of fuel? Could he arrange annual higher purchase payments to coincide with your harvest returns? Did your wife like him? All important considerations indeed!

To add to the dilemma, in 1960 there were no fewer than 22 makes of tractors available in Australia, which accounted for a total of 126 models!  A lot of brochures to pour over!

But at least, away back in the 'good old days' a farmer didn't have to concern himself with the design of the cab or the efficiency of the air conditioning.

# THE START OF THE DAY.

The plum job today on a farm is driving a tractor. Reclining in decibel controlled air conditioned luxury, surrounded by push button and finger tip doodahs for everything, plus a fridge for the Diet Coke and Willie Nelson gently wafting from the quad speaker digital sound system – is not hard to take.

Yes – tractor driving sure beats spending the day working on a fence line or (worse) being ensconced in the farm office endeavouring to entangle the harrowing complexities of the GST and annual tax returns.

The easiest part of driving a modern tractor is starting the engine. A turn of the key - that is all it takes! No need to be concerned with the tedious technological reasons *why* it will start. All that matters is – it *will* start.

But starting a tractor was not always that easy. Particularly if the tractor happened to be a Lanz or KL Bulldog!

## THE BULLDOG.

Starting a semi diesel single cylinder valveless 2 stroke Bulldog tractor was always an interesting way of commencing the farming day. In fact, the procedure if witnessed today at a classic tractor rally, never fails to amaze the spectators who stare wide eyed in disbelief at what they are seeing.

At harvest time, if a Bulldog driver wished to avoid being emasculated by a pack of incensed neighbours, he always stopped his tractor, at the close of the day's work, in the centre of an adjacent gravel road. Country folk driving home from the pub on a dark night had to keep a sharp lookout for parked Bulldogs littering the road. This necessary vigilance was an incentive to avoid nodding off at the wheel.

The reason for this strange practice of parking the Bulldogs out on the road (also applied to the similar HSCS and McDonald tractors of the era) was that in order to fire up their engines – *that is precisely what occurred!* - a roaring flame had to be applied to the cylinder head in order that the tractor could be started. The flame was provided by a highly dangerous and cantankerous petrol fuelled blow lamp. When initially igniting the blow lamp it frequently became engulfed in flames. Therefore it would have been irresponsible not to park these tractors on a gravel surface well free from any inflammable materials, and certainly not in a tinder dry paddock at harvest time.

A blowlamp (sometimes referred to as a "blowtorch" by those recent folk who weren't around in the 1940s) was a gadget much favoured by painters years ago for burning paint off woodwork. (Beats me why they had to do this — but they routinely did just that.) The painter's blowlamp was a toy compared to a farmer's man-size Bulldog lamp.

Half a pint of petrol was poured into the pressure tank at the bottom of the blowlamp and then it was necessary to pump air into the tank in order that the petrol be pressurised. This was achieved with the inbuilt hand pump, providing its leather bucket seal was in good order — which it habitually was not! If such was the case it was then necessary to dismantle the whole thing and fit a new seal — providing you had one! All *very* time consuming.

Theoretically, at any rate, following around twenty vigorous pumps, the tank was pressurised and the valve was opened thus permitting petrol to dribble out of the jet and run down into a pan located half way up the vertical stem between the tank and the jet. By this time the blowlamp was dripping with petrol, as were the hands of the operator. So a deal of courage was required to strike a match and throw it at the lamp. Immediately it would be engulfed in flames. (Now you can see *why* all this had to occur on the road and not in the stubble.)

After a while the conflagration settled down and only the petrol in the pan continued to burn. The resultant heat was required to warm the pressurised petrol enabling it to vaporise. After a few more minutes the valve was again opened, but this time heated petrol vapour was released which burped into an intensely hot roaring blue flame.

So far so good. Hopefully at this stage, by some fluke, the operator had not received multiple burns and only around ten minutes had frittered away. Next, the fearsome blowlamp had to be grasped at *arm's length* and attached to a bracket located beneath a section of the horizontal engine cylinder head, known as the hot bulb. Encouraged by more vigorous pumps, the searing flame was left to its own devices whilst it did its best to heat the hot bulb until it turned cherry red. This would take some minutes, providing the jet didn't clog with specks of dirt in the petrol – a not infrequent occurrence - in which case the entire blowlamp procedure had to be gone through again.

With smoke and flames now erupting under the hot bulb, all according to plan, this was no reason for the operator to pull out his pouch and manufacture a roll-your-own Champion Ruby. There was still much to be done before the inert tractor engine could be started.

Firstly, the big engine had to have its innards lubricated so that when it did eventually start, all its bearings would be nicely dripping with oil. This in fact was an excellent feature not commonly found in tractors with conventional engines. There was however a downside. A hand crank had to be inserted into a fitting in the tractor's lubricating pump and rotated 180 times at a maximum speed of 60 turns per minute, according to the instruction manual. After which, the grease pump (always messy – its normal state) had to be applied to the countless nipples.

Sounds easy enough, except that I suspect the chief Bulldog design engineer had a psychopathic thing against tractor drivers. He had achieved great success in locating nipples in hidden recesses that required a keen memory and the agility of a contortionist to locate. Should the ground below the tractor be wet and muddy, then it was a great start to the day to roll around in the mud under the tractor, wrestling with the greasy slippery grease gun, mindful of the consequences of missing one nipple.

In the meantime, the blowlamp required some extra pumps and inevitably the grease gun would run out of grease half way

*The steering column is inserted into the end of the crankshaft. (Photo M. Daw)*

through the job. To recharge the gun from the bulk can was always an unpleasant and yucky operation. One did not shake hands with a stranger immediately following this particular task.

Let us assume the operator was a responsible tractorman and had filled the fuel tank with crude oil the previous evening. (Never leave a tractor standing with an empty fuel tank!) The task of filling the tank would have involved him in about two hundred wearisome up and down strokes of the hand pump and a likelihood of having firstly to wrestle the 44 gallon drum into position. In the event he had been short of fuel he could have used the sump oil drained from the old Dodge ute a month ago. Bulldogs relished sump oil mixed with the black crude oil. This was definitely the cheapest priced tractor fuel

*Lubricating the grease nipples can be challenging! (Photo M. Daw)*

known to mankind!

A quick check of the hot bulb was now required to ascertain if it had turned cherry red. Prolonged heating might result in a deposit of sizzling molten metal below where the hot bulb should have been!

## THE STEERING WHEEL.

The next step in the starting procedure involved clambering up into the driver's platform, reaching down to the bottom of the steering column and releasing the spring lock. This enabled the entire steering column, complete with steering wheel, to be removed. Clutching this awkwardly cumbersome apparatus the operator, with difficulty, alighted from the tractor and took his burden to the off-side of the tractor. The end of the steering column was then inserted into a fitting located in the end of the east/west crankshaft.

(A passing by non-Bulldog type might have wondered about the unusual spectacle of a steering wheel sticking out from the *side* of a tractor)!

Having achieved the foregoing successfully and with the blowlamp starting to lose its aggressive roar, owing to the petrol therein nearing depletion, the tractor was finally ready for starting.

First, the fuel had to be primed using the appropriate hand lever. Then it was necessary to embrace the steering wheel projecting from the side of the Bulldog and rotate it, one way then the other, with a continuing pendulum motion against the compression created by the huge single piston.  Although only 4.5 to 1 compression ratio, the swept volume of the single cylinder was a massive 10.4 litres.  Consequently to rotate the crankshaft and move the 9 inch piston back and forwards took some effort.

## THE START – BUT?

A few hearty muscle straining swings inevitably coaxed the engine to almost magically cough and burst into a thumping rhythm, at the same time emitting blasts of dense smoke from the chimney.  Horses in adjacent paddocks bolted and galahs screamed their way aloft in protest against the shattering of the hitherto tranquility. Finally the steering mechanism was returned to its rightful place and the blowlamp extinguished prior to being returned to its cubby hole under the bonnet.

It only remained for the operator to climb laboriously, but presumably gratefully, up to his seat and engage a forward gear in preparation to moving off into the paddock. *Stone the bleedin' crows – the tractor is going backwards. The engine is running in reverse cycle.*  Being a valveless two stroke there was nothing to stop it doing just that.

With a bit of skill and luck, the fuel could be cut off at the fuel pump and at the point of stalling, by rapidly priming more fuel, the engine might swing over to run in the correct rotation.

But if the engine had lost some of its pre-heating it would likely be too cold to fire on the rebound.  So the entire starting procedure had to be repeated, beginning with the blowlamp!

A stoical but optimistic Bulldog operator might well have dreamt that someday, someone may design a tractor engine that can be started by *simply turning a key.*

*The same KL Bulldog following its restoration by the author.*

# KILTS, DUNG AND HARRY FERGUSON.

I recently became involved in a discussion concerning Scottish agriculture. Having been raised in a farming community in the ancient Kingdom of Fife, I was able to hold centre stage and bore everyone to distraction as I waxed on about Clydesdales, Golden Wonder potatoes, Model N Fordson tractors, Land Army girls - and dung! Surprising though it may seem, of all of these, dung was (and still is) the most important. Believe me!

For all those enlightened travelling Aussies, who intend visiting Bonnie Scotland, I feel it is my duty to elaborate within these pages the intellectually challenging subject of Scottish dung. Hopefully this will encourage tourists to perhaps by-pass a distillery or two, in their anxiety to discover more about the mysteries of this excellent Scottish product – dung!

A first time intending visitor to Scotland is likely to imagine that all Highlanders wear kilts, have hairy legs, and run hairy looking cattle with dangerous looking horns.

On the subject of kilts, during a recent visit to Edinburgh, the only kilted gentleman I observed parading along Princes

Street was a Japanese tourist!  And the hairy looking cattle are of the Highland breed which require their double layer of shaggy hair to enable them to remain outside throughout the year and graze the glens and moors, even in the bleakest of winters.  I shall not degrade myself by referring to a Highland gentleman's legs, or indeed his kilt or the lack of (or otherwise) associated undergarments!

But I digress.

Scottish farms are the most efficient in the world if measured by yield per hectare continually over a period of many hundreds of years.  Think about that, and the question will follow – why is this so?  Three basic reasons.

## THE ICE AGE.

Back in the mists of time, Scotland was engulfed in an ice age which brought with it glaciers from the Arctic North.  Over thousands of years the ice moved inexorably south, gouging out deep ravines and carrying with it billions of hectares of soil.  The big thaw eventually came at a time when the ice had advanced south to a lateral line, roughly extending from Glasgow in the west to Forfar in the east. The result of the thaw was to dump this enormous quantity of mulchy top soil onto the area, known in geological terms as the Central Lowland Rift.  Today we know the area as Ayrshire, Berwickshire, The Lothians, Fife and part of Perthshire.  The

autumn months, they had stood at floor level.  Following months of the daily addition of straw bedding, the entire "floor" level would have risen to a height of possibly six feet and been trodden into hard packed dung.

Until the late 1940s, forking out the compressed dung, using a forked implement known as a "grape", was undoubtedly the most back breaking job on the farm. Sweating labourers forked out the dung and loaded it into horse carts.  The two wheeled tipping carts, pulled by powerful Clydesdales, were trundled usually some distance to an area in a field where the dung was tipped out to eventually form a long continuing pile of maybe fifty metres.  Progressively, the carts with fresh loads were pulled over the first layer and so on until a rectangular midden of rich dung stood shoulder high in the field.

There it remained, steaming, gurgling and settling for some months.  During this time it enriched as billions of microbes went to work. An agricultural boffin once explained to me that the number of microbes that could be placed upon a threepenny bit was equal to the entire population of people in the world.  And that was fifty years ago!  (For those newer people who are unfamiliar with a threepenny bit – ask your Grandpa).

The following spring the dung was ripe and ready.  The midden was attacked by the same sweating men and horses and carted out and spread into the fields, where deep growing root crops (sugar beet, mangles, fodder turnips, etc.) would be sown.  As these root crops matured they forced the soil apart, admitting oxygen and the precious dung, which enriched the soil in a manner that could not be equalled by artificial fertilisers.  The root crops would be followed the next year by grain, then the field would be put down to pasture for a few years, until the cropping cycle rotation with the dung began all over again.

This method of soil husbandry in Scotland has been handed down by untold generations of farming families.  Today the soil is rich in hummus, worms and the necessary bacteria to yield crops that we here in Australia, with only a scant covering of top soil, can only dream about.

Following a slow start, the mechanisation of Scottish agriculture leapt ahead during the latter part of the 1940s. Aged Model F and N Fordson tractors were replaced by the Fordson Major, these and David Browns, smart in their Hunting Pink livery, became familiar sights across the rural landscape. But it was the little grey Ferguson that, by the early 1950s, dominated the Scottish tractor scene.  Scores of different implements were available for mounting to its revolutionary 3 point linkage system, the brain child of Harry Ferguson and now a standard fitting on most of the world's farm tractors.

But of particular interest to Scottish cattlemen were the Ferguson Manure Loader and Ferguson Spreader. Their introduction instantly relieved farm labourers of the back-breaking task of forking dung out of the byres and steadings and then from the middens.

Suddenly a single man could handle the entire operation without ever leaving the seat of the little Ferguson, and with exerting no more energy than that required to flick a lever.

The spreader/trailer would be uncoupled from the tractor, using the patented Ferguson automatic hook, and parked outside the entrance to the cattle steading. The tractor was then free to enter the doorway and commence forking into the packed dung and emerging with a loaded bucket, the content of which was tipped into the spreader. When the spreader was full, the tractor backed-up to the automatic hitch, hooked up the trailer and hauled the load, at four times the speed of a horse cart, to the field. There the dung was spread in amounts ranging from 6 to 30 tons per acre across the field. All this – without the operator leaving his seat!

Agriculture is big business in Scotland today. Vast improvements in mechanisation, coupled to new strains of hybrid seed, enable this small country, consisting of geographically largely non–arable land, to produce food and fibre far beyond the expectations of its limited acres. But all the modern advances would mean little, if the age old practice of feeding dung to the soil was not still a pivotal part of Scottish farm husbandry.

And now I issue a stern message!

It is my fervent hope that Aussie tourists visiting Scotland will have the necessary fortitude to reduce their time spent on golf, lengthy haggis luncheons, and the sampling of rare malts, and instead apply their energies and intellect to the far more important matter of Scottish dung!

# HARD WORK IN AMERICA.

Over the years Margery and I have clocked up umpteen miles exploring the USA but always doing the tourist thing. In 1997 however it was going to be all hard work. I was under contract to write my third tractor book, to be entitled *The World of Classic Tractors*. By necessity this would see us returning to America for the purpose of carrying out extensive research, involving several weeks of travel across numerous states.

## JAMESTOWN.

We collected our rental car from Hertz at the Nashville airport in Tennessee. The young guy behind the counter wanted to inflict upon us a conservative Volvo. Maybe he thought we looked like Volvo people! But I worked out you don't go to the States to drive a Volvo. So we ended up roaring off in a somewhat ostentatious bright red Thunderbird.

I pointed the big car's nose north east, heading in the direction of the softly undulating Blue Grass country of Kentucky. The rich grazing paddocks on either side of the sweeping roads were garnished with gleaming thoroughbreds, luxuriating in the shade of large spreading chestnut trees. As we drove by, we could not be failed to be impressed by the magnificence of the white gabled homesteads gracing each property.

Eventually we came to the Ohio River and crossed into the State of Ohio. Our destination was the little obscure town of Jamestown where we had arranged to call on Dan Ehlerding, one of the districts prominent farmers, who just happened to own a fine collection of classic American tractors.

The jewel in Dan's collection was a 1916 Aultman Taylor 30-

*The Aultman Taylor 30-60 is powered by a massive 4 cyl. engine which, despite its specification of a 9 x 7 inch bore and stroke, produced only 30 drawbar and 60 engine h.p. The photo shows the author wrestling with the heavy steering, achieved by chains to the axle from the steering box.*

60.  This giant 12 ton tractor stood 12 feet tall to the top of its exhaust stack and had a width of 9 feet.  Its 4 cylinder east/west engine featured a whopping 9 inch bore and a 7 inch stroke.  At its maximum r.p.m. it produced 60 belt horse power and could rattle the machine along a road at a blistering 2.2 m.p.h. A halcyon period was spent becoming familiar with the monster before rumbling it around the trails on Dan's grain farm.

In 1916 most of the prairie giant tractors were unreliable and usually quite dangerous. Crude single and twin cylinder engines were the norm. The Aultman Taylor 30-60 was one of the few exceptions.  Its massive 4 cylinder engine was easy to start and seldom caused problems.  Old timers to whom I have spoken and who had actually worked a 30-60, were full of nostalgic praise for these big machines.

I also had the opportunity of examining Dan's rare 1915 Huber 15-30. Despite the Huber Co. of Marion, Ohio, having previously released five relatively successful models, the 15-30 was in fact an unreliable and poorly conceived tractor. The heart of the problem was the twin cylinder engine manufactured by Sintz Wallen of Grand Rapids, Michigan. The longitudinally horizontally opposed cylinders were fed by a primitive single carburettor located at the top end of cylinder number two.  Accordingly, the petrol/air mixture had to travel the full length of the engine for number one cylinder to fire. This caused erratic running and the inability of the engine to be fine tuned. In addition, the leather cone clutch over-heated and was inclined not to release. A dangerous situation indeed!

Dan was also the owner of a little known 1942 Clark Airborne. During World War 2 Clark Equipment Co. of Buchanan, Michigan, was contracted to build a lightweight crawler tractor fitted with a dozer blade. It was required to be capable of being transported within the confines of a Hamilcar glider, the type used by the Allies for towing behind either anorak Handley Page Halifax or a U.S. Boeing B17. The gliders were released behind enemy lines and upon landing the Clark would be driven out and put to work carving out a rough landing field, thus providing a forward base for fighter aircraft.

To the best of my knowledge only one example of a Clark Airborne is in existence in Australia. It is part of The Puls family collection located near Dubbo, NSW.

We also checked out Dan's 1918 Emerson Brantingham Model Q with its totally open-to-the-elements gear final drives! Curiously, owing to the uncommonly narrow rear width, the front axle had to be extra wide, enabling the offside front wheel to run in the furrow when ploughing, thus keeping the tractor level. This was essential as the 20 h.p. petrol engine relied on splash feed from the sump for its lubrication and this could only be accomplished reliably providing the tractor remained level.

# THE WALLIS BEAR.

We bade farewell to Dan and headed off, this time to another Ohio farming community named Blufftown, where we arrived the next day. My research had alerted me to the fact that a Mr. E. F. Schmidt of Blufftown was the owner of one of the planet's most rare and unique veteran tractors. It was a 1911 Wallis Bear serial number 3.

Classic tractor collectors the world over are invariably friendly and welcoming to fellow enthusiasts. The elderly Mr. Schmidt was no exception. He invited us to follow him in his car from his residence to the outskirts of town where he had a large lock-up shed, in which he housed the Wallis Bear, plus a variety of around a dozen other Wallis models.

The Wallis Bear was conceived in 1902, but the tractor certainly was not rushed into production as it is known that of the nine units produced, number 3 was not manufactured until 1911.

This was another huge tractor! The 4 cylinder engine displacement had a capacity of 24.25 litres. The piston diameter was even greater than that of Dan Ehlerding's machine. The bore of the Bear was 7.5 inches and the stroke 9 inches. Plus, quite incredulously, it was equipped with power steering, independent turning brakes, a spring loaded clutch, enclosed 3 speed transmission, all-speed governor and force feed lubrication by gear-driven oil pump. All this – *designed in*

*1902!* The rear driving wheels had a diameter of 7 feet and a width of 30 inches. Yes, this was *some* tractor!

For the record, in 1927 the Canadian farm machinery giant Massey Harris purchased the J.I. Case *Plow Company*, which then owned Wallis tractors. Thus the first Massey Harris tractors were in fact rebadged Wallis units. (Note; in order to avoid confusion it should be explained that Case tractors were manufactured by J. I. Case *Threshing Machine Company*).

It was hard to tear myself away from this extraordinary artefact, but I was obliged to push on as our next stop was scheduled to be The Henry Ford Museum at Dearborn, Michigan.

*The Wallis Bear, despite its unorthodox appearance, was undoubtedly the most technically advanced tractor during the first decade of the twentieth century. Only 9 units were produced between 1902 and 1911.*

# THE FORD MUSEUM.

Let me make it quite clear, a visit to The Henry Ford Museum in my opinion is simply an unforgettable experience. But for me it was accompanied by a feeling of disappointment.

For some weeks I had been corresponding with a senior member of the Museum team. He assured me he would greet me on the agreed date and personally escort me around the museum's twelve acres. Further, he indicated the book for which I was researching, would help promote the museum and therefore he would render every assistance. For this epistle I shall name him Bob.

I arrived punctually on time on the due date, complete with camera, tape recorder and Margery, who writes things down. It was Bob's day off! No, he had not advised anyone else of my pending visit. No, there was nobody available to escort us around. No, they had no knowledge of my pending book, and it wouldn't have made any difference anyway.

I felt let down, but Margery gave me a lolly which brightened my mood and she took my arm and propelled me into the vastness of the great museum.

We spent a full day roaming the engrossing displays and I shot off around twelve rolls of film (pre-digital era).

A few weeks later, back in Oz, I wrote to the museum offering praise for the quality of the displays and artefacts. I also mentioned that I intended dedicating a chapter to the

Museum, liberally illustrated with a selection of my photos and I would be happy to send them a gratis copy of *The World of Classic Tractors* upon publication.

A fortnight later I received a reply written in aggressive legalistic language stating that I would be subjected to legal proceedings if I failed to first forward for the museum's approval, copies of the photos I intended inserting, to be accompanied with a substantial payment for each photo (I cannot recall the precise amount), plus copies of the associated text, again to be vetted for their approval. But that is not all! Prior to the finished book being released I was required to forward two copies, apparently so they could check I had adhered to whatever stipulations they had imposed! Wow!

I could perhaps be forgiven for consigning the parcel of documents and demands to the waste paper bin. Possibly understandable, but somewhat pettish perhaps, I dismissed any thoughts of including any reference to the Henry Ford Museum in *The World of Classic Tractors*. A shame really because it is a place every tractor enthusiast should strive to visit.

But we had to push on. The next day, we made a brief but constructive visitation to the headquarters of The American Society of Agricultural Engineers (ASEA) their premises being located at the resort town of St. Joseph, perched on the shores of Lake Michigan. From their records and with the assistance of a charming lady, I obtained a deal of valuable historic information.

Our next scheduled visit was to be in Indiana.

## SHEPPARD DIESELS.

Lynn Klingaman's impressive collection of Sheppard Diesel tractors, is located on his rolling grain farm in Northern Indiana, a twenty minute drive from Columbia City. Few Australians are acquainted with these well engineered tractors, which were manufactured at Hanover, Pennsylvania during the late 1940s until the mid-1950s.

Interestingly, during World War 2 the R.H. Sheppard Company designed an ingenious diminutive high revving single cylinder diesel engine with its own encapsulated fuel system, specifically for powering ships lifeboats. The little engine could run for a duration of 18 hours whilst coping with the rigors of stormy seas and continue running even if upside down! Thousands of these engines were supplied to the U.S. Navy.

In 1949, following extensive testing, Sheppard Diesel tractors were introduced to Mid-West American farmers. Contrary to the normal industry practice of releasing one new model at a time, Sheppard unveiled no less than three distinct examples each with a number of available variations, including row-crop and high clearance models.

It should be noted that the correct name of the Sheppard

produced tractors is *Sheppard Diesel.* The tractors were never sold as simply *Sheppard.* Accordingly, it is incorrect not use a capital D in connection with the *Sheppard Diesel* range of tractors.

The top selling tractor in the Sheppard Diesel stable was the model SD3. As with all others in the range, it was powered by Sheppard's own design diesel engine, in this instance a 3 cylinder unit developing 32 h.p.

The flag ship of the range was the big handsome SD4, which came complete with power steering, disc brakes and an advanced engine monitoring system. The 319 cu. inch 4 cylinder engine, produced 51 belt h.p. at a leisurely 1650 r.p.m. A 10 forward speed transmission was standard, but a farmer could choose to have his SD4 equipped with a torque convertor

*Lynn Klingaman, whose farm is located in Northern Indiana, driving a magnificently engineered Sheppard Diesel SD4.*

which still retained the 10 speed transmission and provided a torque multiplication ratio of 2.14 occurring at 1540 r.p.m.

Sheppard was one of the few tractor manufacturers capable of designing its own diesel injection system. The inline pumps were incorporated in a single housing, but could be easily individually calibrated by a farmer having only limited mechanical knowledge.

The injectors had only a one pin-hole orifice and no return pipe. This enabled a diverse range of fuels to be utilized, including low priced furnace oil. It also required the higher than average compression ratio of 22.5 to 1.

Lynne proved to be a generous host and insisted I drive each of his tractors and later his cheery wife provided a scrumptious lunch served on the scrubbed kitchen table. When it was finally time to depart, our hosts presented us with Sheppard Diesel caps and T-shirts.

## THE DUPLEX CO-OP.

Morristown is south east of Indianapolis. The highway from Columbia City goes via Fort Wayne and is a half day drive from Lynne Klingaman's property. It was now nearly a week since we had commenced our drive from Nashville.

Jack Cochran and his son Mark were waiting for us as we drove up to their homestead. Lined up in precision order were four beautifully presented Co-Op tractors.

Around 1934, distinguished engineer Dent Parrett (the co-owner of the Parrott Tractor Co. of Chicago) was approached by the Farmers Union Central Exchange Co-operative of Minnesota to design and have manufactured a range of tractors, specifically for sale to their Co-op members. Appropriately the tractors would be named Co-Op.  Parrott arranged for Duplex Machinery Co. to Build the Co-Op units at their plant at Battle Creek, Michigan.

In 1936 the first of the range was released – the Co-Op 1. This was followed over the next two years by the Co-Op 2 and 3.  The numerical digits indicated the number of plough bottoms (mould boards) each tractor could comfortably pull.

The Co-Op 1 was powered by a 4 cylinder 16 h.p. Waukesha engine, whilst the 2 was equipped with a 30 h.p. 201 cu.

*The author takes the wheel of a CO-OP 3, powered by a 40 h.p. 242 cu. inch Chrysler engine. (Photo M. Daw)*

inch Dodge side valve unit and the 3 by a 40 h.p. 242 cu. inch smooth running Chrysler engine.

It was evident by the immensely strong channel chassis that these were extremely rugged tractors. The engine was set low in the frame and coupled by a tail shaft to a 5 speed Clark gearbox and a Dodge truck rear transmission.

The final Parrott designed Co-Op was the S3 and did not appear until 1948. It retained the Chrysler engine but now featured a down draught carburettor and disk brakes.

In 1949 an entirely new range of Co-Ops appeared, but they were merely re-badged versions of Canadian Cockshutt tractors and, according to Jack Cochran, lacked the character and appeal of the original Co-Ops.

I discovered that driving Jack's Co-Ops could be likened to driving a truck. They were fast (*very* fast) but easy to control owing to their low centre of gravity.

Jack then insisted we visit a near neighbour named Charlie Schilling, the owner of a 1930 Massey Harris General Purpose 4 WD, plus a strange looking 3 wheeled 1916 Case 10-20. Darkness was closing in as we arrived at the property, therefore disappointingly I had only a brief opportunity of inspecting Charlie's two artefacts.

The Massey Harris was a model with which I was relatively familiar having operated one in New Zealand. This was the first tractor designed from the ground up by Massey Harris, the earlier examples being virtually rebadged Wallis units. Somewhat euphemistically the tractor was marketed as a high clearance row crop model. The Achilles heel of the tractor was the placing of the heavy 24 h.p. Hercules engine forward of the front driving axle. Steering the machine required the muscles of a navvy! Crudely designed turning brakes were of little assistance. Having steel lugged wheels was a tedious problem if the tractor was to be driven on a bitumen road, as four bands had to be fitted to the wheels, as distinct from two with a conventional two wheel drive tractor.

Whilst the odd ball Case 10-20 was known to me, Charlie Schilling's was the first example I had ever sighted.

Remarkably only the offside furrow rear wheel was a driver, unless a dog clutch was engaged in which case both rears became drivers. However, as there was no differential, it was imperative that the tractor only proceed in a straight line, when the two wheel drive clutch was engaged! The tractor's redeeming feature was a modern overhead valve 20 h.p. engine. This was liquid cooled by a radiator precariously mounted sideways alongside the single front wheel.

## JOHN DEERE HOSPITALITY.

The sun was setting low in the west as we bade farewell to the Cochrans. After skirting Indianapolis, Lafayette and Chicago, we then drove west across Illinois to Moline. Overnight stops were no problem as the American Central West has modern motel clusters on the fringes of most towns.  Diners and restaurants are also everywhere.  Our two favourites were Cracker Barrel, with their wood fired ovens and juicy gourmet steaks, and Red Lobster with their scrumptious Alaskan shellfish.

The following forenoon we arrived at the Deere & Co. Corporate Headquarters on the outskirts of Moline where we were warmly greeted by Rollie Henkes, one of their senior marketing executives, who by arrangement was expecting us. Despite obviously having a full and heavy workload, we were

*The unit pictured is a weird looking Case 10-20 dating back to 1916. Although having 3 wheels, only the furrow wheel was a driver, that is unless the near side rear wheel was engaged, by means of a dog clutch. However, if both rear wheels were engaged in the drive position the tractor could only proceed in a straight line, as there was no differential. The overhead valve 20 h.p. 4 cyl. engine was liquid cooled utilising a radiator mounted sideways alongside the single front wheel. This example was restored by Charlie Schilling.*

assured he was more than happy to devote as much time with us as we could afford.

I noticed Rollie had both my previous tractor books on his office shelf and he indicated his interest in my current project - *The World of Classic Tractors.*

When I expressed a desire to include a feature chapter on John Deere in the book, Rollie emphasized the fact that his Company would gratefully appreciate this and added "Each time the words John Deere are in print, that represents free advertising for us. We are proud of the name".

Rollie guided us through the museum portion of the elaborate building and then escorted us through the climate controlled arboretum with its tall greenery. He explained the remarkable concept of the piped 'silence' that promoted a feeling of calmness within the office complex. At lunchtime he treated us to an excellent repast in the corporate dining room. He also arranged an appointment for us the next morning with Les J. Stegh, the Deere and Co. Senior Archivist.

Les and his staff occupied an upper floor of the John Deere harvest assembly building, overlooking a wide bend of the Mississippi River, on the western outskirts of Moline. They had recently completed the mammoth task of filing several tons of documents and records relating back to the year 1837, the year that the young John Deere invented the self-polishing steel plough.

The key to the workings of the system were explained to me and I was encouraged to spend as much time as I required in obtaining whatever information I desired. Also his staff members were there to assist if necessary.

Curiously there were no records of a visit to the John Deere factory by the German farm machinery industrialist Heinrich Lanz in 1902. This, despite the fact that I have inspected the relative documents held in the Lanz Archive building at Mannheim, which for over half a century has been part of the German John Deere complex.

The Mannheim documents confirmed that Heinrich Lanz had been corresponding with John Deere and exchanging ideas of technology for some years up until the death of John Deere in 1886. But it was not until 1902 that Lanz undertook the arduous journey to Moline where he was welcomed by Charles Deere, the son of the founder.

Lanz discovered that the Moline factory could boast double the output per man hour compared with the Mannheim plant. On his return to Germany he introduced a programme of modernisation in accordance with the John Deere practices. (Heinrich Lanz A.G. later went on to produce the legendary Lanz Bulldog tractor).

But despite this omission in the Moline archives, there was an abundance of valuable material for my book, including some rare never-before-published photos.

I shall long remember the kindness and generosity of the folk Margery and I met at Moline.

## THE GALLOWAY.

We left Moline behind and crossed the Mississippi into Iowa, that vast terrain of flat endless acres of cornfields stretching beyond the horizon in all directions. Here I had to remain extra vigilant and constantly monitor the speedometer to make sure the 70 m.p.h. speed limit was not exceeded. State Troopers, Highway Patrol and Deputy Sheriffs in their Ford Victoria cruisers seemed to be around every corner.

The Sheriffs were particularly aggressive. If you were in their particular county and caught speeding then whoopee – your fine contributed to their wages. On the one occasion I was pulled over for exceeding the speed limit by a Highway Patrol officer (in Arizona) all he wanted to do was talk about Australia. His only rebuke was a mild warning to drive safely "...as you never know where these pesky sheriffs will be hiding" he cautioned.

We paused on our way through Cedar Rapids for fuel, then further on we exited the highway as we approached Waterloo and took a back road leading to the farming district of Dunkerton. This was the location of Kenny Kass and his incredible collection of very early prairie tractors.

During the long bleak Iowan winter months whilst his

extensive corn fields lie dormant, Kenny Kass specialises in the restoration of Waterloo Boy tractors. His reputation for excellence is known among tractor enthusiasts around the world. Indeed at least two Australian collectors have each imported superbly restored Waterloo Boys from Dunkerton.

Whilst I was keen to study his methods of restoration, this was not the main attraction for me. I had come to inspect Kenny's Galloway Farmobile, possibly the world's sole remaining example. So following several hours of admiring his restoration craftsmanship and blissfully clambering up onto his extensive range of huge early prairie tractors, it was time to fire up the Galloway.

In 1916 the Wm. Galloway Co. of Waterloo, Iowa, agents for Cadillac automobiles, branched out into the business of manufacturing farm tractors. Certainly Charles William Galloway and his wife Hellen knew they would be faced with competition from other nearby pioneer tractor makers. Within a two days buggy ride there was The Rock Island Plow Co., The Waterloo Gas engine Co., plus The Heider Mfg. Co., each successfully manufacturing and marketing tractors to an increasing number of mid-western farmers who were prepared to embrace the new technology.

The Galloway Farmobile was powered by a 4 cylinder 20 h.p. side valve engine manufactured by The Dart Truck and Tractor Corp. originally of Anderson, Indiana, but which Galloway purchased in 1910 and re-located to Waterloo. The tractor was equipped with two forward and one reverse gear.

Apparently Galloway received a contract from Henry Garner, a British car dealer in Birmingham, for the supply of a quantity of Galloway Farmobiles to England, during that country's shortage of tractors during the 1914-18 war. It appears (although I have been unable to determine the accuracy of this) the tractors never left their port of embarkation owing to the prevailing wartime shipping priorities. Certainly no payment was received by Galloway, which contributed to Wm. Galloway Co. going bankrupt in 1919.

The Galloway proved recalcitrant and uncooperative that bitterly cold winter day and the engine refused to fire. So Kenny and I were obliged to push it out of its place of retirement into the daylight, in order for me to take the necessary photos. But to me, just clapping eyes on such an exquisite and rare piece of ironmongery was a joy in itself.

## AN EXPERIENCE IN THE SNOW.

The cold was bitter as we continued our way heading west along the Iowa State Highway 20. The endless cornfields retained their winter mantle and the landscape was grey under a sullen sky. The darkness was setting in, yet it was only 4 p.m. when we thankfully checked in to a welcoming motel, situated just off the ring road around Fort Dodge.

It was a case of 'on the road again' the next morning. The Thunderbird's heater was on full blast, as the outside temperature was well below zero. There was an eerie stillness and the sky had dawned a menacing purple grey. Being born and reared in Scotland, I was familiar with the signals. Snow, and lots of it, was imminent!

However we were obliged to soldier on, as we had appointments to keep. Our next stop was scheduled to be a farm near the township of Pomeroy, a one hour drive from Fort Dodge.

Pomeroy turned out to consist of a single street, boasting two used tractor lots and a line of tired looking clapboard houses. The only citizen to be seen was an old guy on a veranda wrapped in layers of coats and scarves and who could have stepped right out of a Hemingway novel. He was ensconced in a rocking chair, apparently watching the traffic going by, except ours was possibly the first car through Pomeroy that day!

The farm to which we were heading was around a fifteen minute drive north of the town and was the home of Steve and Rachael Rosenboom and their two teenage sons. The purpose of our visit was to inspect and experience the alleged world's fastest production tractor.

This may seem to some as being a somewhat frivolous item of research. But not so. The tractor in question was the brain child of a couple of brothers whose name happened to be Friday. They operated a small agricultural machinery manufacturing business in the Michigan town of Hartford. In the aftermath of World War 2 there was a huge demand for tractors out on the vast mid-west grain belt and indeed tractor dealers usually had a lengthy waiting list for new customers. Sensing an opportunity, the brothers designed and produced an innovative tractor which, not surprisingly, they named the Friday.

The first batch of Fridays, introduced into the market in 1948, were powered by the readily available Ford V8 engines, but this was soon changed to Chrysler 6 cylinder 218 cu. inch units, developing 100 brake h.p. In 1948 this rendered the Friday a seriously powerful tractor. By design (or oversight) the big engines were not fitted with a governor! The gear train consisted of a 2 stage 10 speed transmission. In top gear, with the Chrysler engine revving out to a mere 2,500 r.p.m., the tractor would theoretically be low flying at 90 m.p.h. *Theoretically* is the operating word here as presumably the unsprung tractor would have been uncontrollable.

Never-the-less, Steve told me that he had recently pulled away from a pickup truck travelling at 60 m.p.h. Steve is a good Christian guy and I am sure he was telling the truth. He admitted to having to hold on and utter up a wee prayer as the Friday tore along in front of the pickup. From 1949, all new

Friday tractor engines were equipped with governors. (Possibly at the behest of some spoil-sport accountant or legal advisor).

The Rosenbooms ushered us into the farm kitchen where a distinctly 'different' vegetarian lunch was laid out before us. (A whole boiled cauliflower – and I do not like cauliflower! But it was kind of them and *'manners maketh man'* my father used to say. So I tucked in, hopefully giving the impression I was enjoying a great delicacy).

We sat around the table and yarned for around two hours. The Rosenboom boys I think were relieved that we Aussies could speak fairly good English! In the meantime it had started snowing, gentle flakes at first and then the sky opened up and all vision outside the windows became totally opaque. It eased by mid-afternoon, enabling us to waddle through a foot of pristine snow to Steve's shed in which was housed the Friday.

The bright red low slung machine to me had similar characteristics to the Offenhauser speedway racer with which Ray Revell used to thrill crowds at the Sydney Showground Speedway. Well – maybe not *quite* like that, but when the big engine was fired up, the sound emitted from the short stubby straight-through exhaust pipe was definitely reminiscent of the racket one associates with speedways.

But what now? We simply had to be in Lincoln, Nebraska the following morning, yet the road was now carpeted with a foot of snow. Rachael kindly offered to accommodate us until the weather cleared.  But unfortunately our arrangements at Lincoln meant this was not an option.

"If you get going now whilst there is still some daylight left, you should make it to the Interstate Six, where the snow ploughs will have been busy".  Steve's advice was not totally reassuring, but we had no option but to get going. The boys shovelled away the snow which had banked up against our car.

Armed with photos and information relating to the Friday, we waved farewell and headed for the road.  But the snow had levelled everything, including the four foot deep ditch that ran adjacent to the exit to the road, and into it plunged the front of the Thunderbird. What a nightmare!  Well not really, because in a matter of minutes Steve had the Friday connected to the rear of the stricken car attached by a long rope.  In no time the car was extracted unharmed from its predicament.  The only collateral damage was to my self esteem.

The Thunderbird slithered its way somewhat dramatically for the hundred mile drive to the Interstate Six.  We drove in darkness and nearly lost control a number of times. However driving in snow and ice was not a new experience for me, but I have to confess to being thankful when we reached the highway, which had been swept of snow by the time we pulled off the treacherous secondary road.

We were both fairly weary and decided to overnight at Omaha, just across the state border into Nebraska.  The first motel we sighted seemed good to us. After checking in we noticed that opposite in a blaze of neon lights was Jack's Joint. By an unspoken mutual agreement it was decided that Jack's Joint required our immediate investigation and a sampling of that establishment's liberally proportioned New York steaks - well done with fried onions on the side!

## THE NEBRASKA TEST.

All classic tractor enthusiasts are familiar with the Nebraska Tests. During the pioneering days of tractor development, a number of unprincipled American manufacturers enticed unsuspecting farmers to purchase tractors that were devoid of engineering integrity. These machines were frequently dangerous and inherently unreliable.

One such company operated by W.B.Ewing, deceptively marketed his tractors as a 'Ford'. He simply traded on the famous brand name without having any connection whatsoever with the legitimate Henry Ford operation. He is also alleged to have employed an illiterate youth named Paul Ford to put his signature on the tractor blue prints in an endeavour to further hoodwink prospective buyers.

The tractor was an engineering disgrace but unfortunately for Ewing, one was purchased in 1916 by a Nebraska State Legislator named Congressman W.F. Crozier of Polk County. His Ford tractor simply was incapable of performing any normal tractor tasks. For twelve months Crozier was subjected to promises and excuses by Ewing, but without results. In frustration Crozier purchased a second tractor – a Big Bull from the Bull Tractor Co., of Minneapolis. The Bull proved to be no more reliable than the Ford. Both the Big Bull and Ford were powered by a Gile horizontally opposed twin cylinder engine, a product of The Gile Boat and Tractor Co., of Ludington, Michigan. The engine was designed for powering motor boats—no less!

Crozier discovered that many farmers were experiencing similar problems with their tractors. Something had to be done!

In association with Senator Charles Wentworth of Nebraska, Congressman Crozier successfully introduced new legislation into Nebraska law in 1919 that required all tractors offered for sale in Nebraska to be first rigorously tested by The University of Nebraska Agriculture Engineering department. Tractors failing the test would be prevented from being sold in that state.

The new legislation had a profound effect on the entire tractor industry. Nebraska was a major sales outlet for tractors and no legitimate manufacturer could risk the disgrace of having a tractor rejected by the test authority. Accordingly the shonky manufacturers largely disappeared from the scene and the remainder made certain that their tractors were thoroughly prototype tested prior to being introduced into the market.

We navigated our way through the maze of Lincoln City and eventually were met at the main entrance to The Tractor Testing Laboratory of The University of Nebraska by a friendly Brent Sampson, the senior engineer of the facility. Brent proudly explained the accomplishments and history of the establishment, which had been in operation continually since 1920, apart from the years of Word War 2. I owe Brent a debt of gratitude for the amount of valuable information I obtained from him during our visitation.

## THE MAIL ORDER TRACTOR.

The following morning we called on Vern and Grace Anderson, retired farmers now residing in Lincoln. Vern's shed contained his perfectly preserved 1938 Graham Bradley row crop tractorman big sleek unit reminiscent of the stylish American cars of the period. Set low under the long bonnet lay a 6 cylinder Continental side valve 217 cu. inch motor. There were 4 forward gears, including a 22 m.p.h. transport gear and by over riding the governor this speed could be increased to around 40 m.p.h. Interestingly, tractors in Nebraska were much cheaper to register than trucks, therefore farmers preferred not to own a truck and instead transport their products to market on trailers hauled behind high speed tractors.

A fascinating aspect of the Graham Bradley is the fact that it could only be purchased by mail order. Many of the American homesteads were remote from shops and tractor dealers in the years prior to the post war era and before the construction of the Federal interstate highways. Mail order catalogues were a way of life. Sears Roebuck was the leading mail order supplier and offered everything from a shovel to a homestead, which would be delivered to the nearest rail siding following the receipt of a ten percent deposit. Included within the pages of their catalogue was the Graham Bradley tractor.

Farmers were given five years to pay-off the competitively priced tractor at attractive low interest rates, plus a plough and a cultivator were included in the deal at no extra cost!

So how good was the Graham Bradley? The tractor was

*Vern Anderson, of Lincoln City, Nebraska, is shown astride his sleek Graham Bradley row crop tractor. Although produced in limited quantities, the Graham Bradley proved to be a well engineered and reliable tractor.*

quality all through, infinitely comparable with the big name manufacturers such as John Deere, Case, Oliver, International, etc. The majority of sales however were restricted to farmers in Indiana, Illinois, Iowa and Nebraska.

It was quite an experience driving Vern's tractor around his estate. My impression was of silky smoothness, enhanced by the single plate semi-centrifugal patented Velvet-grip clutch.

## BACK TO NASHVILLE.

Having concluded our scheduled visitations it was now time to wind our way back to Nashville and return the rental car to the Hertz depot. We had enjoyed three absorbing weeks and would spend another seven days or so meandering back along an entirely different track, calling in to agricultural museums along the way in the expectation of acquiring additional historical facts that would be useful in future publications.

Our route embraced Kansas, Oklahoma, Missouri, Arkansas, and back across the Mississippi at Memphis into Tennessee and on to Nashville.

Following the month long research, I felt completely satisfied and relaxed with the results. When publishing historical information concerning tractors, so far as I am concerned its accuracy is sacrosanct. Regrettably a number of young writers today without any farming or tractor background, noting the saleability of tractor books, are cobbling together facts true or often false, which they have gleaned from a variety of sources. They surround them with glossy photos and have them published and offered to an unsuspecting public as gospel information.

# HOFHERR-SCHRANTZ-CLAYTON-SHUTTLEWORTH.

To most unenlightened non-tractor folk, the above name is probably suggestive of either one of these incomprehensible menu selections with which one is confronted in a typical yuppie Little Collins Street restaurant, or some weird cocktail also available in one of the aforementioned establishments.

But to those of us on a higher intellectual plane, i.e. blessed with a knowledge of old tractors, we know of course that the HSCS is in fact the name of a quirky Hungarian tractor.

It is not surprising that with such a mouthful as Hofherr-Schrantz-Clayton-Shuttleworth, the company preferred to be known as HSCS.

## HSCS IN THE BEGINNING.

The origins of HSCS extend back to the 19th. Century and combined British and Hungarian interests – an unlikely combination at that time.

Nathaniel Clayton was a crusty old English sea captain, but also an acknowledged expert in the field of steam engines. In 1842 he retired from his navel career and formed a business partnership with Joseph Shuttleworth – a respected engineer. They established a firm in Lincolnshire for the purpose of designing and manufacturing marine engines. However in response to the rapid expansion of mechanised farming, the factory switched to the production of agricultural steam

*The Booleroo Steam and Traction Preservation Society Inc. of South Australia are custodians of this fine example of a 1937 H.S.C.S. L25 crawler tractor, which has been painstakingly restored by Ferg Innes. The large chimney is typical of single cylinder semi-diesel 2 stroke engined tractors. Owing to their design, during combustion not all the fuel is burnt, therefore the oily smoky residue would immediately clog a normal exhaust pipe.*

engines together with a range of portable threshing machines.

Clayton and Shuttleworth Ltd. experienced a dramatic growth of business and became one of the largest and most prestigious steam engine manufacturers of the period and ranked alongside such firms as Heinrich Lanz of Germany, J.I. Case of the USA. and Marshall and Sons of England.

By the year 1862, Clayton and Shuttleworth were exporting their machines from Lincolnshire to eager markets in Continental Europe as well as Russia, Africa and The Far East.

## BUDAPEST.

Towards the end of the 19th Century, a group of wealthy Hungarian Magyar noblemen, who jointly owned vast tracks of Hungarian prairie farm land, had ambitions of establishing a farm machinery manufacturing plant in the nation's capital Budapest.

The principal share holders in the proposal were Matthius Hofherr and Janos Schrantz, both agricultural engineers of distinction. A decision was taken to approach Clayton and Shuttleworth with an offer for them to purchase an interest in the new Budapest enterprise, thus bringing with them their expertise and engineering skills. The Englishmen enthusiastically embraced the proposal and thus in 1900 the firm of Hofherr-Schrantz-Clayton-Shuttleworth came into being. The first units to bear the brand of HSCS were a range of state-of-the-art threshing machines.

The Hungarian manufacturing facility prospered and diversified. In 1921 the first HSCS tractor was introduced. The German firm of Heinrich Lanz A.G. had offered the services of its chief tractor design engineer Doctor Fritz Huber, to assist with the design of the new tractor. It featured a single cylinder hot bulb type engine, which was ideally suited to farmers with no prior mechanical experience.

The 2 stroke valveless configuration of the HSCS engine was simple and reliable. Almost any type of combustible fuel could be used, including lamp oil, used cooking fat and old sump oil drained from other vehicles. Starting the engine required the pre-heating of the hot bulb section of the cylinder head with a blow torch, prior to rocking the flywheel in a pendulum motion until the engine fired into life, generally accompanied by a shattering explosion and a spectacular emission of sparks from the tractor chimney.

## AUSTRALIA.

In 1934 Western Australia was the first state to receive an importation of HSCS tractors. This consisted of three K50 'Steel Horse' models. In total 80 HSCS units were sold to Western Australia farmers prior to the outbreak of World War 2. Wheat farmers in that state relished the fact that a 'Steel Horse' would consume only 12 gallons of fuel whilst performing hard work during a full eight hour day.

In 1949 following an absence of ten years, HSCS tractors (now Hungarian Communist state owned) were re-introduced into Australia by Brown and Dureau Ltd. The most popular of the range, which included crawler tractors, proved to be the G35. In actual fact there were very few technical differences between the 1949 tractors and the original 1921 models. The transmission, brakes, steering and cosmetics had been improved over the years, but the engine remained basically the same single cylinder hot bulb design.

By the mid 1950s, Australian farmers were tending to become weary of the noisy and bone-jarring characteristics of the single cylinder design. The smoother running multi-cylinder engines with their push button starts, were certainly more user-friendly and desirable than the single cylinder 2 stroke designs.

But those farmers who still remained dedicated to that reliable and simplistic concept, found that the Lanz Bulldog company had modernised their single cylinder tractors to the extent that they were now eminently more sophisticated than the HSCS machines. Accordingly, Lanz sales in Australia remained healthy, whilst HSCS tractor sales fell away until the tractor was quietly removed from the market during the mid 1950s.

Interestingly, in 1955 the Hofherr Schrantz Clayton Shuttleworth company was renamed Dutra. It appears the

Communist State was anxious to obliterate all references to a Capitalist past. The factory switched its production priority to that of heavy dumpers and also trucks. In 1961 a range of heavy duty four wheel drive Dutra tractors was introduced, and were among the earliest true heavyweight tractors of the post war era.

During the 1970s Dutra amalgamated with another state owned firm named Raga and a licensing arrangement was entered into with the Steiger Tractor Company of Fargo, North Dakota. The Budapest company was renamed Raga Steiger and produced a range of Steiger tractors largely for the East European market.

# THE CHAMBERLAIN SUPER 90

## ONE OF THE WORLD'S ALL TIME GREAT TRACTORS – AND IT WAS AN AUSSIE!

## THE EARLY CHAMBERLAIN TRACTORS.

The Western Australia based Chamberlain Industries Limited entered the tractor scene in 1949 and, during a somewhat turbulent first decade, produced a range of excellent tractors eminently suited to broadacre grain fields.

The first Chamberlain tractor offered to Australia's farmers was the 40K, powered by the firm's own twin cylinder, transversely mounted, horizontally opposed 40 h.p. engine designed to operate on low priced power kerosene. Not only was the engine unique in design, but went against the global trend of equipping tractors with inline multi-cylinder power units. Even John Deere, Lanz and Marshall were about to abandon the concept of their idiosyncratic single and twin cylinder engines.

Two years following the introduction of the 40K, the 40KA became available. The two models were virtually identical, but the 40KA incorporated a closer range of gear speeds in its nine speed gearbox.

The tractors were true heavyweights, tipping the scales at 8,500 lbs. They proved popular, particularly in Western Australia. However by the time the model was superseded in 1955, actual sales amounted to only 2,000 in total, well short of the 1,000 units per annum that had been initially envisaged.

The final production run of the twin cylinder Chamberlains occurred in 1955 when a small number of 55D tractors were introduced. They featured the same configured engine but with a clever but complex designed combustion system, coupled to a variable compression ratio arrangement, enabling them to run on diesel fuel following a cold start on petrol. (It is interesting to note that this starting procedure was successfully used by International Harvester in their first diesel engined tractors).

In 1955 the Welshpool factory surprised the industry with the release of its Champion range of tractors, powered by the indestructible Perkins 4 cylinder 270D diesel engine. An imaginative promotional exercise, saw a slightly modified Chamberlain Champion farm tractor serving as the recovery vehicle for the 1955 round Australia Redex Trial, in which it recorded 17,824 kilometres in 19 days. The Champion became the sales leader in its class and the most profitable tractor in the Chamberlain stable.

Protected by a tariff bounty system, Chamberlain tractors were priced extremely competitively against imported machines. But as the American manufacturers introduced increased sophistication and larger engine capacities into their tractors, Chamberlain was obliged to keep pace.

Accordingly, a prototype 60DA incorporating a 3 cylinder General Motors 2 stroke diesel was trialled, but it was found that its 60 h.p. could not match the drawbar pull of the emerging Oliver, John Deere, Minneapolis Moline and other American heavyweights. Thus the Super 70, a near identical unit to the 60DA but with the G.M. diesel boosted to 70 h.p., was rushed into production and the 60DA relegated to the history books. But even the Super 70 was still lacking in performance when compared to its American competitors.

## THE SUPER 90.

In 1963 Chamberlain released the Super 90, a restyled unit tipping the scales when water ballasted to around 7 tonnes and featuring the legendary G.M. 371 power plant. Originally configured to produce 90 h.p. a new updated version of basically the same engine delivered 100 h.p. at 1,800 r.p.m. was used in the Mark 2 version of the Super 90. The 213 cu. inch capacity at first glance seemed inadequate in relation to its performance. However the 2 stroke engine with its three cylinders was supercharged (as distinct from turbo charged) and the Super 90 was capable of exerting a massive drawbar pull of 8,460 lbs. at 3.1 m.p.h.

By way of comparison the Oliver 1900D, the most powerful of the American imported tractors in the early 1960s, under test at Nebraska returned a 12,475 lbs. drawbar pull but at the much slower speed of 1.83 m.p.h. and in order to achieve traction it carried an additional 4,740 lbs. of ballast in the form of stacked weights clamped to the rear axles. At 4.95 m.p.h.

*One of the world's most powerful 2 wheel drive tractors of the era, the Chamberlain Super 90, was introduced in 1963. The original G.M. 90 h.p. 371 supercharged 2 stroke 3 cylinder unit engine was up-graded to the 100 h.p. version. The modified tractor was identifiable by a new design of a one piece beam front axle, but retaining the soft riding transverse leaf springs. The beefy 23.1 x 26 inch rear tyres, mounted on heavy cast steel rims and with water ballast, tipped the scales of the Super 90 to around 7 tonnes. This enabled the tractor to transmit all its horsepower to the ground, producing an impressive drawbar pull in excess of 8,500 lbs at 3.1 m.p.h. This example has been restored by the author.*

the pull of the Oliver was reduced to 6,282 lbs.

The figures can be confusing, but the important factor here is that the Super 90, which in normal operating trim weighed around 1,500 lbs. more than the Oliver, did not require extra ballast to efficiently transfer its power to the ground. An interesting element in this comparison is the fact that the Oliver 1900D was powered by a G.M. 4 cylinder 2 stroke super charged engine of 212 cu. inch capacity and the Chamberlain Super 90 had the G.M. 3 cylinder 2 stroke super charged engine of 213 cu. inch capacity.

The bulky broad section 23.1 x 26 rear tyres of the Chamberlain, mounted on heavy duty cast steel rims, plus the sheer weight of the tractor, enabled the Super 90 to obtain its phenomenal grip of the ground. In terms of pure grunt, the 100 h.p. Super 90 was possibly the most powerful two wheel drive farm tractor in its era. Indeed farm implement manufacturers at that time were unable to manufacture ploughs and scarifiers of sufficient size to match the available power of the Super 90. Therefore it was a common sight when driving through Australia's broadacre country to see a Chamberlain Super 90 working comfortably with two 22 disc ploughs coupled together in tandem.

The Super 90 had 9 well spaced gears and a hand operated over centre clutch which was a joy to use. The offset upholstered bucket seat provided a relaxed driving position and was perfectly positioned to enable the operator to obtain a clear relatively unobstructed view ahead. The front axle had a soft leaf spring suspension, which rendered the tractor uncommonly soft riding and therefore less fatiguing when long days had to be endured.

The high cost of manufacturing the Super 90 proved uneconomical in relation to a realistic retail price and therefore created severe budgetary problems for the Company. As a consequence it was phased out in 1966 and eventually replaced by the Countryman 354, powered by a Perkins 6–354 engine. Certainly more modern in its appearance and in its own way a good solid tractor, but the Countryman was no substitute for the Super 90.

The Super 90 was a classic in its own time. Older generation farmers when discussing their "pride and joy" tend to get a glazed expression as they nostalgically recount the many incidents which endeared the Super 90 to their hearts forever, despite the fact that today they may well farm with a state-of-the-art 4 wheel drive 300 h.p. air-conditioned computerised powerhouse.

# ROMANIA.

Of all the iron fisted Communist regimes in Eastern Europe prior to 1990, the evil malignant rule of Nicolae Ceausescu in Romania was by far the most feared and despised. It is now over two decades since a well placed bullet penetrated his upper cranium, putting an abrupt end to Ceausescu's reign of terror and debauchery.

I experienced at first hand the "delights" of communist Romania when, in pursuit of tractor knowledge, I visited that land of nightmares just months prior to the people's uprising in 1989.

## BREAKING THE RULES.

I broke all the rules when, accompanied by Margery, we entered Romania by the "back door". Western businessmen were supposed to fly into Bucharest and be met by a member of the State Foreign Trade Department, who would serve as an "escort" during the restricted visitation, until being returned to the airport.

Accordingly, the Frontier Guards were confused and unsure how to react to our unscheduled arrival by car, at a remote border crossing. So their instinctive reaction was to do what came most easy to them – shout and point Kalashnikov assault machine guns in the direction of our vehicle.

The day was damp and gloomy. We had just crossed the Danube from Bulgaria at a place called Ruse, having motored through Bosnia, Serbia and Bulgaria to the Black Sea coast. Our car was a rental Zastawa, which nobody (including us) had ever heard of hitherto. The unshaven scruffy Romanian Frontier Guards were accustomed to receiving Bulgars, minor officials and Communist Party delegates arriving from Ruse. Suddenly, there were we with our Zastawa, sporting Serbian Belgrade number plates and being driven by an Australian resident brandishing a UK passport! All very confusing!

A large uniformed oaf bellowed at us to get out of the car. We hastily obliged. His fellow hoodlums, with their large nicotine stained fingers, then proceeded to paw their way through our suitcases containing our personal possessions. In the meantime, another goon commenced pulling the trim from the doors of the Zastawa and the cushions from the seats. Were they looking for guns, drugs or simply being objectionable? A junior officer in a stained uniform stood contemplating us whilst he picked and examined the contents of his nose.

I'd had enough! I shouted at *him*! He blinked in astonishment. I extracted a document from my briefcase and displayed it before him. The official looking paper had been presented to me in Sydney by the Romanian Consulate and was adorned with impressive looking seals. Written in Romanian, it presumably stated that we should be treated as VIPs and shown courtesy and respect.

The nose picking individual took one glance at the paper and rushed off to the adjacent guard hut. A moment later he re-emerged, followed by a tall swarthy obviously senior officer, to whom he pointed us out. The new comer adjusted his tie and cap then saluted and welcomed us to his fun loving and affable country.

## THE SECURITATTI.

Having navigated the now re-assembled and re-packed Zastawa a mere 100 metres into Romanian territory, we were again ordered to stop – this time alongside a decidedly upmarket guard hut. A long lean tight-skulled looking chap, in an ankle length black leather coat and black hat, approached the car. It was obvious he had seen all the World War 2 Gestapo movies and dressed accordingly. He had apparently been radioed ahead by the frontier guards, as we were greeted politely in English. This was our first of many encounters with the Securitatti (State Secret Police).

Around half an hour was spent scrutinising our passports, our official document and the accommodation reservations at Brasov. He informed us, courteously but firmly, that his Securitatti colleagues would expect us to arrive at our hotel no later than by late afternoon. His tone indicated this was not a suggestion but a warning to comply with his directive. He also gravely told us that it was "forbidden" to visit any residence in Romania. Nor should we under any circumstances take photos of public buildings, military installations and personnel, roads and bridges or factories, anywhere in Romania. Additionally it was not permitted to photograph *anything* within 10 kilometres of the frontier, which included the Danube River. I figured out that it might just be ok to photograph a cow under a tree in the centre of Romania.

## THE DRIVE TO BRASOV.

The speed limit throughout Romania varied depending upon the class of vehicle. I was disenchanted to learn that our Zastawa would be restricted to a maximum of 80 kilometres per hour but pleased that I was not driving a 4 wheel drive, all of which were restricted to 60 kilometres per hour. A breach of speed regulations by a Westerner would result in experiencing the ambience of a Securitatti cell. Consequently as we ambled along the main road to the capital Bucharest there were no difficulties in threading our car through the horse traffic, which constituted 50% of all road vehicles. The remainder were either official black Russian built Lada limousines, decrepit looking Romanian made Dacia 1300 S sedans, or tractors.

The tractors were exclusively UTB (Uzina Tractorul Brasov) variants hauling up to three wagons loaded with a variety of goods ranging from farm produce, oil drums, coal, and scrap iron to groups of conscripted peasant labourers with downcast eyes. The tractors were fitted with air compressors, required to actuate the air brakes with which the trailers were equipped. It appeared that in Romania, these tractor trains were frequently used in place of semi-trailers.

Although April, the wide farmlands on either side of the road to Bucharest were still clad in their dull grey winter mantle. However in some fields UTB tractors were opening up stubble country with mouldboard ploughs, often several tractors to fields of around twenty acres.

I was soon to learn that 99% of all tractors employed in Romania were UTBs produced at the giant UTB factory at Brasov, to which we were journeying. Despite driving around 1,500 kilometres within Romania and observing hundred of tractors at work, I only saw five units that were foreign imports. These were ageing Soviet Belarus machines, purchased at a time when Romania was still friendly with Soviet Russia.

Bucharest was a one hour drive north from our border crossing. As we had failed to notify The Department of Foreign Trade of our intending arrival and had no accommodation pre-booked in Bucharest, we were not permitted a stop over in that city.  Margery navigated from a map on her knees as I drove the Zastawa through the one time stately capital and found the road north to Brasov.

During the 170 kilometre journey from Bucharest to the Medieval city of Brasov, our attention was diverted from our usual practice of observing the passing farmlands.  Positioned each 200 metres, during the first one hour of travel north of Bucharest, were groups of either two or three Kalashnikov bearing soldiers.  There they stood, like forgotten monuments looking miserable and decidedly damp in the dismal afternoon drizzle.

It appeared that the previous week, factory workers from around the nation had marched upon the capital in a courageously defiant but futile protest against the acute food shortages which threatened the health of the Romanian workers and their families.  The armed troops along the way constituted a show of force by the authorities and a deterrent against further rebellion.

In addition to the armed soldiers, we also encountered elevated Securitatti check points each ten kilometres.  Signs indicated that the car had to be slowed to 15 k.p.h. as it was driven past, permitting our progress to be monitored and reported.  On each occasion binoculars were focused upon us by officials high in their towers.

We motored through bleak villages and noted that many of

the side roads, around which the cottages clustered, were of compacted soil and undoubtedly would become transformed into rutted mires in wet weather. Outside a few of the cottages were Dacia cars, generally elevated on blocks of timber. The supply of petrol to ordinary citizens had been cut off since the previous autumn.

Further, we were to learn that electricity, even during the bitter winter months, was not permitted to be used by households between 10 p.m. and 5 a.m. There were but two hours of television broadcasting each evening, comprising largely of government propaganda. Yet, at the approach to each village was a large signboard proclaiming the benevolence of Nicolae Ceausescu and his love for his Romanian citizens. The bullet through his skull some months later suggested that not all Romanians returned his love!

## BRASOV.

This remarkably preserved ancient city is located in a valley deep in the Carpathian mountains. In 1925 an aircraft factory was established on its outskirts. The location had been selected owing to the bountiful supply of ash timber from the adjacent forests. Ash was an extremely flexible but strong timber used in the manufacture of aircraft frames up to and throughout World War 2.

During the German occupation of Brasov in World War 2,

the factory was commandeered by the Third Reich for the production of the formidable Junkers JU 87 Stuka dive bombers. It was considered that the remoteness of its location would protect the plant from Allied and Russian air attacks.

At the end of the war Romania became a Soviet Communist satellite state. The factory was transformed into a tractor production facility. East German tractor technicians, who had little choice in the matter, were recruited into the running of the factory. The first tractor emerged from the Uzina Tractorul Brasov factory in 1946. It was a 4 cylinder 30 h.p. petrol engined machine – the Model IAR 22 – with design

*An aerial view of the vast Uzina Tractorul Brasov manufacturing plant located on the outskirts of Brasov. The sight originally was an aircraft manufacturing factory and a close scrutiny of the photograph reveals aircraft hangers at the far side of the complex. Junkers JU 87 Stuka dive bombers were produced here during the German occupation in World War 2. (IMJ archives)*

characteristics reminiscent of an American 1920s tractor.

Had it not been for the grim atmosphere that prevailed and the resulting furtiveness of its inhabitants, Brasov would have been a tourist's delight. Strolling through the narrow winding streets, which meandered between ancient buildings, was indeed a joy. There was also a quiet serenity about the place, partly created no doubt by the lack of vehicular traffic in the old sector.

But the shops were barren of fresh meat and fresh vegetables. Curiously however there was an abundance of bottled gherkins. Bread was strictly rationed and long queues formed wherever it was available.

The Hotel Carpathi at Brasov had obviously been a building of considerable architectural beauty in the 1920s and 30s. In 1988 it reposed in a state of decaying elegance, like an impoverished ageing baroness. Its vast foyer was illuminated by six feeble light bulbs. In its days of glory the pillars and ornately carved domed ceiling would have been ablaze with light from the thousand glistening crystals of the magnificent chandeliers, now hanging like abandoned parasols coated in dust and cobwebs.

Positioned inside the revolving entrance doorway day and night, stood a uniformed soldier brandishing a Kalashnikov. Nice friendly touch I thought. During our four-day stay I made a deliberate attempt to extract a smile or nod from whichever soldier happened to be on duty, but without luck!

Two factory officials visited us at the Hotel during the evening of our arrival. Unlike the curt manner of the officialdom we had come to expect, the factory representatives greeted us cordially and we warmed to them immediately. They each spoke excellent English which was a welcome relief following my lame attempts at the Romanian language,

They informed us apologetically that the sole factory car had been despatched to Bucharest for a few days and would I mind if they arrived by trolley bus the following morning and then used our car for the drive to the factory. I had to make a mighty effort to disguise my amazement that one of the world's largest tractor factories possessed only one car! Was this the true face of Socialism?

The Hotel Carpathi was reserved for Communist Party officials and an assortment of delegates mainly from East Germany, Bulgaria and Poland. The crumbling edifice was also the obligatory hotel for the occasional visiting Westerners having business at the UTB tractor factory.

Breakfast in the one-time elegant dining room offered limited choices but would have been considered a banquet by the seriously under nourished citizens of Brasov, where gaunt eyed children were growing up never having tasted butter or fresh milk and seldom experienced meat or fresh fruit. Even The Carpathi substituted yoghurt for milk, rendering my breakfast cup of tea – well, *different*!

The two representatives from the factory, whom we had met and warmed towards the previous evening, arrived at the hotel at 8 a.m. having travelled by trolley bus. Their names were Dan and Adriana. Again they apologised for having to request we use our Belgrade rental Zastawa for transport instead of the one (and only) factory car that had apparently been despatched to Bucharest.

The eight kilometre route from the hotel to the factory ascended a gradual climb out of the valley in which Brasov was centred. As we left the old part of the city behind I steered the car through street corridors of ugly high density flats, so typical of the Communist reconstruction of Eastern Europe during the 1950s and 60s. Margery and I found the presence of the grim faced armed militia occupying most street corners, somewhat unsettling.

## THE FACTORY.

The UTB (Uzina Tractorul Brasov) complex is a sprawling mass of buildings inter-connected by a network of service roads. Not surprisingly the site is totally flat, having once served as the airfield adjacent to the still remaining original aircraft factory structures.

I parked the car as directed by Dan outside the brick and glass fronted administration block, close to the main entrance.

Margery then extracted my cameras from the car boot. This brought an exclamation of scandalised dismay from Adriana, who insisted they be returned to the car out of sight, at once! Dan covertly glanced at the always present armed guards to determine if they had noticed the offending cameras. He then nodded to a large sign adjacent to our car which emphatically stated that cameras were "Forbidden" in the area outside or inside the factory.

Being asked (told) to leave my cameras behind when on tractor safari, was equivalent to being asked to remove my trousers. I felt naked! Being a tractor scribe necessitates the taking of photos to illustrate my articles. An appeal to the factory manager later, to be allowed to take limited photos within the factory, was abruptly dismissed with a wave of his hand.

Dan and Adriana ushered us into a stark oak panelled reception office. The main item of furniture was a well worn conference table around which were a number of bolt-upright chairs. Four senior department managers shuffled into the room accompanied by a halo of tobacco smoke. Following introductions we all took our seats around the table. Mugs of scalding instant coffee (with no milk and tasting like brewed iron filings) were banged in front of us by a stern faced woman with the physique of a rugby league centre-forward.

The executives around the table smiled cordially at us and

each in turn broke into a lengthy monologue. I am not sure why they bothered because at the completion of their half hour performance, during which we could not understand one word being uttered, Dan translated and encapsulated all that had been said. This took him around three minutes! The subject matter had been mainly government propaganda, however we did learn that tractors from the Brasov plant in 1989 were being exported to 83 countries world wide.

I interrupted the proceedings to visit the "Gents" and was directed to a door at the far end of a long corridor. With the door firmly closed behind me, I stared with astonishment at the toilet paper. The mottled brown cardboardy textured material had been made with wood shavings, which were clearly visible! I had learned by now that certain commodities in Romania were fairly crude but this was, I thought, a bit rough!

Accompanied by our two guides and the factory manager, we then spent a fascinating few hours strolling through the immense complex, 50 hectares of which, we were told, were under roof. This was no cutting edge high tech. factory but certainly appeared functional and efficient. The workers in the main were industrious and many of them flashed us a smile, to be abruptly turned off as they noted the Securitatti officer trailing along behind.

I took a careful look at the tractors destined for Australia being assembled in a special quality control area. This was the result of pressure brought to bear by Joe Jardin of Inlon Pty. Ltd. (the importer of UTB tractors into Australia) upon the Romanian Foreign Trade Minister, who had personally arranged this special treatment by the factory's best technicians specifically for the Australian bound tractors.

## THE NEW AND THE OLD.

At precisely 3 p.m. we were conducted to an open compound in which a prototype of the new UTB (Farmliner) 3 Series was presented for my inspection and approval. The new tractor certainly featured pleasing modern lines, plus a number of technical improvements had been incorporated. Under the skin I was pleased to note that it was basically the same solid but unremarkable UTB, however with noticeable refinements in the areas of castings and machining.

I suggested nine areas that in my opinion required further development for the Australian market. I felt my comments were appreciated and obviously heeded, as the modifications were incorporated into the Series 3 tractors which arrived in Australia some months later.

The afternoon terminated with a visit to the factory museum and, being a tractor historian, the highlight of my day was being given a drive of the 1946 first ever UTB produced. Frankly it was reminiscent of a 1920s style tractor, therefore to me – fascinating.

The next day was spent further inspecting the UTB plant, however the third day turned out to be *decidedly* different!

## DRACULA.

Following breakfast, we set off from the hotel in Brasov crammed into our Zastawa, with Dan and Adriana the self appointed navigators. Our destination was Bran (Dracula's) Castle – a two hour journey high into the Transylvanian Alps.

Everyone has read, or at least knows of, Bram Stoker's classic book "Dracula". Very creepy and not to be read alone at night. But few realise that Dracula was a real person.

Tractors were forgotten as Margery and I listened enthralled to Adriana's informed account of Dracula's history, as we climbed ever deeper into the mysterious Transylvanian Alps.

Dracula's full title was Count Vlad Tepes Dracula the Impaler, Prince of Wallacia, son of Vlad Dracul, grandson of Mircea the Great, King of Wallacia. (Pretty impressive stuff, huh?)

Dracula ruled Transylvanian from 1448 to 1476 and lived in the fortress citadel Bran Castle to which we were heading.

His title "The Impaler" was both accurate and appropriate. You see, he had this nasty and unpleasant habit of inviting dignitaries, who might just possibly pose a future threat to his rule, to dine at the castle. Having gorged on venison and quaffed liberal quantities of slivovitz, they were then abruptly herded into the courtyard, where Dracula's playful guards awaited them.

One by one each unfortunate dignitary was held and had a three metre stake forced down his throat, until it protruded through his groin. The stakes were then hammered into the ground in rows on either side of the road leading up to the castle, in the manner of fence posts. The grizzly corpses remained on display for twenty days and served as a deterrent to any would be over-ambitious subject.

Adriana went on to tell us that a pair of visiting Turkish noblemen refused to doff their hats to Count Dracula. In response their hats were nailed to their heads with *50 centimetre nails!*

Yes – a nice type indeed and I wasn't sure I really wanted to visit his castle after all.

It was a grey damp mid-morning when we eventually arrived at the mist enshrouded infamous Bran Castle. A more sombre, ugly and unfriendly place would be hard to imagine. The Dracula presence was almost tangible. Margery remained very close to my side as we meandered through the ancient building, exploring the secret passages recessed and honeycombed within the four metre thick walls. I noted we all talked in whispers. It was freezing cold as our footsteps echoed within the many chambers.

# ESCAPE TO FREEDOM – NEARLY!

The following day we departed Brasov. Our route from Brasov west to the Hungarian border passed through some stunningly beautiful mountain scenery, but sadly interspersed with melancholy villages devoid of happiness. We were saddened to observe gaggles of gnarled peasant crones, hurrying to surround the car and pleading for food whenever we stopped adjacent to their habitation. Some of the larger townships down on the plains existed only for their industrial slums, each with belching chimneys breathing vile yellow fumes upon the populous. There was a total absence of greenery in these rancid areas.

We finally exited Romania at a forlorn frontier crossing near the town of Oradea. Kalashnikov armed mounted frontier guards, tall in the saddles of their sweating horses, cantered through the timbered border woods hoping to arrest (or shoot) some wretched person endeavouring to flee the tyranny of Nicolae Ceausescu. There were no regrets therefore as we were grudgingly dismissed by the suspicious Romanian authorities and drove the 200 metres through "no man's land" towards the Hungarian frontier officials. My sigh of relief was replaced by anxiety as the young green-uniformed Hungarian passport officer politely informed me that our visas were not "appropriate", therefore we could not enter Hungary!

But that is another story!

# THE DIESEL DIFFERENCE.

## THE MENACE OF WHITE VANS!

In Britain it is *large* white vans. In France it is *small* white vans. The thing is, it appears to be a pre-requisite that to drive a white van in the land of fish and chips, or across The Channel in the land of *haute couture*, one is required to have suicidal or maniacal tendencies! These vans are everywhere. Their diesel engines scream for mercy as they are thrashed from the moment they are fired up and propelled into Formula One mode by psychopathic nut cases. Now for the record and in defence of blue van drivers or red or even green van drivers, I have to state that they are not a problem.

The remarkable thing about all this is the fact that these high revving diesel engines, flogged in this manner, don't simply go *POOF* and disintegrate. Testimony indeed to the backroom boffins of modern technology.

A few years ago during a visit to Europe, accompanied of course by my Girl Friday – Margery Daw, I was prompted to rent a diesel powered station wagon. Well I mean to say, all the creditable motoring scribes have been writing such glowing reports about the brilliance of these new common-rail diesels being fitted into cars, that I felt it was uncool not to have driven one.

The engine in the wagon was a mere 4 cylinder 2,000 c.c. unit. Despite it being turbo charged, I admit to anticipating it would be a bit of a slug on the road, as the vehicle weighed-in at just over 1.5 tonnes. How wrong I was! With the beautifully matched 6 speed transmission, this wagon was a low flyer!

On the superb autobahns, it happily cruised all day at a comfortable 140 kph whilst recording a lazy 2,200 rpm on the dial. Without the worry of speed limits, and with Speedy Gonzalez Margery urging me on, I confess to giving the vehicle its head on a few occasions and with total safety blistered along at a pace, which in this country, would have had a highway patrol officer thinking all his Christmases had arrived at once and I would have languished in jail for the rest of my days.

But even whilst I was doing my Stirling Moss thing, diesel powered VW Golfs and Skoda Octavias thundered past leaving us in their wake, as if we had been sitting at the side of the road enjoying a picnic.

The good news was that the white van brigade largely steered clear of the autobahns, preferring to inflict their intimidating presence on the winding secondary roads. You know – the ones with blind corners and humpback bridges.

We clocked up around 4,500 kilometres in our wagon and remarkably it returned a miserly fuel consumption of 5.9 litres per 100 ks. This included driving in cities and highways. Pretty impressive stuff!

Of course fuel is more expensive in Europe than in Oz, but significantly diesel is on average 20% cheaper than petrol. (Why is it that Australia is possibly the only country on the planet where one pays more for diesel than unleaded petrol, when it costs much less to produce? I think we all can guess

*The Lanz Bulldog HL, designed by the legendary Dr. Fritz Huber and introduced in 1921, was possibly the first volume selling tractor not fuelled by petrol/kerosene. Indeed the low compression single cylinder two stroke engine could be fuelled with any low octane combustible liquid, including vegetable oil, naphtha, ship's bunker fuel, or even sump oil drained from other vehicles. This example restored by the late Eric Bolwell.*

the answer!)

You can probably sense I am now hooked on diesel engined cars.

My hitherto bias against such vehicles had its origins spawned back in the 1950s, when a Sydney taxi firm put on a fleet of diesel powered Vanguards powered by 4 cylinder Standard engines, borrowed from the Ferguson FE 35 Diesel tractor. These cabs were so gutless that it was not uncommon, when driving up the original, near perpendicular, one way Spit Hill at Seaforth, for passengers to be requested to alight and walk to the summit, before continuing on with their taxi ride.

Even with later car diesel engines, they always sounded as if some errant washers were clattering around within their innards. Also they were smelly!

But not any more! Our European missile diesel engine could hardly be heard from within the vehicle and even when idling the sound was indistinguishable from a well bred petrol unit.

## EARLY DIESEL TRACTORS.

Not surprisingly (being me) all this diesel business had me thinking about the early days of oil burning tractor engines. The technology, then in its infancy, was so far removed from today's diesel tractor engines that it is hard to believe that these modern marvels had their geneses in those far off smoking belching power plants.

During the first decades of the Twentieth Century there were numerous attempts by tractor manufacturers to power their machines with other than petrol engines. Kerosene was a cheap alternative but proved less efficient than petrol. However it could be used in a more-or-less orthodox petrol engine but required the unit to be initially started and warmed on petrol.  Kerosene does not vaporise until heated.

The first non-petrol/kero volume selling production tractor was the German built Lanz Bulldog HL, introduced in 1921.  Its extraordinary but revolutionary 2 stroke design rendered it the ideal machine for the majority of the era's farmers, whose only knowledge of horse power related to the four legged variety and whose technical expertise was limited to sharpening of a scythe and possibly the setting of a mouse trap.

In 1922  Benz Sendling, another German make, released the world's first tractor equipped with a full compression diesel engine.  In hindsight, it is plainly obvious that the chief designer of this astonishingly weird peace of ironmongery must have had a deranged animosity towards tractor drivers.   Yes – it was that bad!

The three wheeled layout featured a single large diameter skeleton rear wheel, driven from the engine by a massive overgrown bicycle chain.  This rendered the centre of gravity of the  monstrosity well above the centre of the rear wheel. Accordingly, if driven on any surface more mountainous than a bowling green, the three tonne machine would fall over onto its side!

But never one to be flummoxed, the Teutonic design geezer appended to the rear of the tractor a set of outrigger wheels, similar to the trainer wheels he had noticed on his grandson's first two wheel pushbike. This at least rendered the machine less likely to capsize and exterminate the operator.

However the designer really needn't have bothered, because an endeavour to fire up the engine generally resulted in failure. Infact one very unhappy owner, who happened to be a South Australian farmer, after finally getting his Benz Sendling started, kept it running night and day during the harvest

*The German firm of Benz Sendling is credited with producing the world's first full compression diesel engine tractor. In addition to the unwieldy 3 wheeled version the company also produced a 4 wheeled configured unit using the same 5.3 litre 32 h.p. 2 cylinder upright engine. This example restored by Norman Bates.*

season, because he could not be certain if or when he could restart the thing if he shut it down!

The farmer elected to withhold his hire purchase payments as a protest aimed at the dealer, Messrs. E. Schrappel of Tanunda, who had talked the farmer into buying the contraption in the first place. In retaliation, Messrs. E. Schrappel sent a technician to the farm to remove the cylinder heads from the engine until payments were resumed. All quite extraordinary!

So what was so bad about the engine? The concept of a full compression ignition was innovative and represented forward thinking. But sadly the design of the Benz Sendling example proved to be under-developed and was prematurely offered to unsuspecting farmers.

The 4 stroke engine featured 2 cylinders in line, each with separate heads. Its 5.3 litre capacity produced an alleged 32 h.p. at a mere 800 r.p.m. To assist starting from cold, two ignition papers had to be lit and inserted into orifices leading into the combustion chambers. This was usually followed by a prayer, the crossing of fingers and toes, then some physically demanding activity on the crank handle. All too often such enterprise only resulted in a defiant inert non-running engine and a thoroughly unhappy farmer!

It is interesting to contemplate that a modern tractor diesel engine, having a capacity of 5.3 litres, generates around 120 h.p. It will likely have 6 cylinders and consume a fraction of the fuel gulped by the 1922 Benz Sendling.   Best of all – it can be started by simply touching a button!

Spare a thought for our pioneer tractor drivers.

And, if in Europe, keep a sharp lookout for these mentally challenged white van drivers.

# THE DARK SECRET.
## A TRACTOR TALE OF INTRIGUE!

## THE JOURNEY BY NIGHT.

The night was pitch black and the lashing rain continued to bucket down. The driver of the old Albion rubbed his tired eyes as he strained to penetrate the feeble glow from the headlamps and endeavoured to avoid the worst of the potholes. His co-driver sitting alongside had dozed off into an uneasy sleep, miraculously, considering the continual jolting of the cabin.

The year was 1928, and it was four hours since the lorry, with its furtive load well hidden beneath a black tarpaulin, had been disembarked off the cross Channel ferry at Dover. And fourteen hours since the journey had begun at Düsseldorf.

To stop for a cuppa was out of the question! The instructions were emphatically clear. The Albion with its cargo *had* to arrive at its destination under the cover of darkness. Then be securely locked away, well before the workers arrived at the plant and turned curious eyes in the direction of the laden factory transporter, wondering why the company logo had been crudely obliterated with black paint.

The first streaks of pink announced the dawn sky, as the Albion was finally navigated into the innards of the custom-built windowless construction, located deep in the interior of the factory complex. With the lorry safely inside, the single steel door was secured with a massive padlock. The drivers were weary and anxious to arrive home to a hearty breakfast and a warm fireside. A fifty pound note was handed to each and a reminder that the continuation of their employment with the Company was totally dependent upon their keeping their mouths tightly shut with regard to the clandestine operation.

In fact, their silence and loyalty were assured. The Great Depression was just getting into its vicious stride in 1928 and jobs were scarce. Further, at a time when two pounds ten shillings per week was considered a generous wage, neither of the men had even *sighted* a fifty pound note, far less received one. The dark secret would remain secure!

## THE MARSHALL COLONIALS.

In 1928 the distinguished Lincolnshire firm of William Marshall Sons & Company, the world's largest producer of steam engines and allied machinery, was in serious financial difficulties. The fact that a respected English organisation, able to trace its origins back to 1848 and with voluminous exports radiating to the four corners of the globe, was now teetering on the edge of bankruptcy, was indeed indicative of the perilous state of

British heavy industry!

Marshall had made an effort to diversify from its steam heritage when in 1906 it unveiled a prototype farm tractor, powered by an internal combustion engine. The power unit was a 2 cylinder petrol engine designed by Herbert Bamber – a highly regarded engineer better known for his association with The Vauxhall Car Company. Bamber's tractor engine featured a hefty 7 x 7 inch bore and stroke.

By 1910 the Marshall factory (The Britannia Iron Works), had developed two tractor engines based on the Bamber prototype. A 2 cylinder version developed 35 brake h.p. and a 4 cylinder unit produced 70 brake h.p.

Known as Colonials, the tractors were massive, indeed the Type G, the largest, weighed 13.25 tonnes. Even the 2 cylinder 'lightweights' weighed in at 8 tonnes. Accordingly, British farmers were not enchanted with the Marshall Colonials as they were simply too heavy for the soft moist tilthy arable soils. A bogged Colonial presented a daunting problem as it could take a team of around twenty heavy draught horses to extricate the stricken machine.

As a consequence the majority of Colonials were sold overseas where they worked on shallow and often hard baked soils.

But there was a further problem! They were besieged by mechanical faults and the volume of sales required to establish an economy of scale did not eventuate.

By the mid 1920s, Marshall's fortunes declined alarmingly. The Colonials had been discontinued with only around 300 produced. The era of steam power was rapidly coming to an end. Farmers around the world were becoming attracted to more efficient lightweight tractors. Fordson, John Deere, International, Saunderson and other relatively inexpensive makes were enjoying rapidly expanding sales.

In Europe an odd-ball tractor known as the Bulldog with a horizontal single cylinder engine, was being manufactured at

*The Marshall Colonial Class E as pictured was powered by a 2 cylinder 35 h.p. engine designed by Herbert Bamber (of Vauxhall car fame). This particular unit, upon its arrival from Britain, was transported by paddle steamer from Adelaide to Menindee - Western NSW, and driven overland to Clare Station. The author is shown driving the big machine through the grounds of Victoria's Swan Hill Pioneer Settlement, where the unit is on display. (Photo M. Daw)*

the giant Mannheim plant of Heinrich Lanz. AG.  Its low cost and simplicity of design encouraged non-mechanically inclined farmers, most of whom had grown up in the horse or mule era, to embrace the new tractor technology and purchase one of these somewhat idiosyncratic Lanz tractors.  By 1930 the Lanz Bulldog had become Europe's top selling tractor, plus a thriving export market had flourished.

## THE CLANDESTINE PLAN.

In 1927 the chairman of William Marshall Sons & Company, Herman Marshall,  grandson of the founder, established a Committee of Investigation, the purpose of which was to produce an innovative plan, guaranteed to reverse the decline in revenue and return the firm to a position of profitability.

Several ideas were advanced by the Committee, but the one most favoured by Herman Marshall and his board was the concept of producing a copy of a Lanz Bulldog.  It was considered the 2 stroke single cylinder Lanz would be easy to reproduce, with minor variations thus avoiding breaches of patents.  Such a tractor it was believed would attract volume sales in Britain and in traditional Marshall export markets.

A plan was drawn up whereby a Lanz Bulldog would be acquired surreptitiously in Germany and stealthily transported to Gainsborough, where it could be dismantled and examined. The Committee stressed that absolute secrecy was essential in order to avoid alerting opposition manufacturers of the project.

A special windowless warehouse was constructed to house the Lanz, enabling design engineer Samuel Dawson to proceed with his measurements and drawings unobserved.

Dawson was by nature an innovator and not entirely happy with the concept of merely creating a virtual copy of the Bulldog. He believed he could considerably improve the design of the single cylinder 2 stroke engine by increasing the compression ratio from 5 to 1 to 15.5 to 1. Thus instead of being a semi-diesel it would become a full compression diesel engine.

From one major aspect, this was a mistake. In actual fact a significant factor relating to the success of the Bulldog was its low compression ratio engine. Being a *semi*-diesel enabled it to be fuelled with a variety of cheap low cost products, including crude naphtha oil, sump oil drained from other vehicles, peanut oil, indeed just about any combustible fluid with a low octane rating.

An additional important factor of the Bulldog low compression engine was its *inability* to burn the total volume of fuel injected into the combustion chamber. Accordingly, there was always a degree of unburnt oil washing the inside of the cylinder wall, thus reducing the friction and therefore wear of the piston, rings and cylinder wall.

But even if some wear did take place, there would be virtually no noticeable fall-off in engine performance.

There was one other major advantage of the Lanz engine, when compared to the full compression ignition design proposed by Dawson. As the fuel was detonated within the combustion chamber of the Bulldog, owing to the low compression, the detonation explosion was not instantaneous. Instead, it 'leisurely' chased the piston to the bottom of its stroke in the manner of a steam engine, thus providing the Bulldog with a substantially higher torque characteristic than tractors with conventional either diesel or petrol engines. In other words – a farmer got a lot more pulling power per horse power with his Lanz Bulldog.

No doubt Dawson considered these factors, but contumaciously pushed ahead with his alternative design. Perhaps being persuaded by the fact that with the full diesel the tedious business of having to first pre-heat the cylinder prior to starting, as in the case of the semi-diesel, was eliminated.

## THE MARSHALL DIESEL TRACTORS.

Samuel Dawson worked secretly and diligently, pushing ahead with the design of his single cylinder diesel engined tractor.

In 1930, the tractor industry and farming community were taken by surprise, when Marshall announced, with considerable fanfare, the introduction of the Marshall 15-30 tractor. However, sadly for the manufacturer, field tests were to prove the new Marshall in terms of performance was no match for its Lanz equivalent. Additionally the fuel injection equipment was unacceptably troublesome.

In 1931, the 15-30 was re-equipped with a German Bosch fuel pump and injector, but still its reliability and performance were considered unsatisfactory. Few were sold and those that had been exported to overseas dealers were returned to the factory at Marshall's expense!

A replacement model, the 18-30, was introduced in 1932. However despite the Company outlaying capital (which it could ill afford) on the development of the new model, there was only a slight improvement with its reliability.

Astonishingly, there were only seventy two Marshall single cylinder tractors sold between 1930 and 1934!

The Marshall 12-20 arrived on the scene in 1935, with a reduced 9 inch cylinder bore and a new cooling system, with Lanz style radiator cells mounted crossways above the cylinder block.

The Marshall Board must have given a collective sigh of relief, for finally Samuel Dawson had got it right! The 12-20 was a creditable tractor, easily started by a hand crank, once the cigarette-like starting igniter had been inserted into the combustion chamber.

But alas the 12-20 had arrived too late to make a worthwhile impression on the British tractor market. A mere 212 were produced.

At the commencement of Word War 2, in 1939 tractor manufacturing at the Britannia Iron Works was switched to the production of war materials. That is apart from a small number of Marshall Model M tractors, ordered by The Ministry of Supply, aimed at assisting farmers to increase their yield of desperately needed farm produce. The model M was a stop-gap, basically a 12-20 with increased engine revs.

At the cessation of hostilities in 1945, Marshall announced a much improved new range of tractors, bearing the name of Field Marshall. But *that* is another story!

*A Marshall 12-20, originally a rusted out derelict when discovered, has been restored by the author.*

# APPENDIX.

The special secret warehouse, into which the Albion lorry transported the Lanz Bulldog that dark night in 1929, was considered so secure and impenetrable from prying eyes, that the British War Department  acquisitioned it in 1940 for the purpose of developing the top secret midget submarines, designated X Craft.

German Intelligence remained unaware of the existence of these 60 feet long undersea stealth vessels until in 1943 the pride of the Germany navy, the 43,000 ton Tirpitz, was blown apart at her moorings in a heavily defended Norwegian Fiord.  The crew of an X Craft had attached limpet mines to the underside of the battleship!

# TRACTORS AND AUSTRALIAN POLITICS OF THE 1940S.

During the Second World War, new tractor availability came to a more-or-less full stop.  Australian farmers were obliged to patch up old tractors that in normal times would have been scrapped.  To add to the problem, spare parts were in a critical short supply, and worn out pneumatic tractor tyres had to be replaced with hitherto abandoned spudded steel wheels. Often horse teams, having been replaced by tractors and now enjoying their retirement in back paddock, were once again harnessed to the plough.

Following the end of hostilities, the shortage of farm tractors continued.  The diminutive indigenous Australian tractor manufacturing industry, which included A. H. McDonald and Co., Howard Auto Cultivators Ltd.,  Ronaldson Bros. & Tippett Ltd., and Jelbart Bros., had either switched to war related production or gone out of business.

In 1944, the directors of the Melbourne based firm of Kelly and Lewis, had commenced talks with the Commonwealth Government regarding the urgency of producing a local tractor for Australian farmers. These discussions related to licences being granted to KL for the purchase of scarce steel and an inducement for the firm to build a new foundry. It was decided that the company would build a copy of the German Lanz Bulldog, for which KL had been agents prior to the war. It would be known as the KL Bulldog.

Owing to a farcical series of bureaucratic government inter-departmental procrastinations, coupled to funding restriction imposed upon the project by the KL accountants (resulting in a walk-out by a totally frustrated Alios Murr, the chief project engineer), the tractor was not released until 1949, two years behind schedule! The predicted production target was to be 1,000 tractors per year. Infact, only 900 of the big single cylinder KL Bulldogs were produced during its five year production life!

In the meantime, the Western Australia state government had been involved in negotiations with Bob Chamberlain and a group of financiers, who wished to establish a tractor manufacturing plant at Welshpool, an outer Perth suburb. Unlike the dilatory performance of their Commonwealth counterparts, the WA government officials eagerly and efficiently facilitated the proposal and agreed to certain generous funding as an incentive to getting the factory established and in business. Chamberlain tractors went on to become one of Australia's top selling tractors until eventually being taken-over by Deere and Co. in 1970.

Back in 1912, the American owned International Harvester Company had established an Australian subsidiary. In 1937, the firm opened a farm machinery manufacturing plant at Corio Bay, near Geelong, Victoria. But it was not until 1948 that new extended premises were opened, enabling the production of International Farmall tractors to commence.

A limited number of British imported tractors arrived towards the end of the 1940s. Included were the Fordson Major, Ferguson, Field Marshall, David Brown and the Scottish built Massey Harris 744D, each enjoying the tariff advantages of The Empire Trade Preference Agreement. A scheme designed to give preferential tariff benefits to British manufactured goods.

Also a trickle of North American tractors, including John Deere, Massey Harris, Case, etc., began once again to be imported, however without the price advantage extended to British manufacturers.

Due to an inexplicable and obviously wrong decision made in 1948 by the Prime Minister, The Rt. Hon. Ben Chifley, Australia missed out in having one of the world's largest tractor producers establish a tractor manufacturing plant in either New South Wales or Victoria.

## ENTER OLIVER.

Today's younger farmers will likely have little or no recollection of the name of Oliver, in relation to farm machinery and tractors. This is not surprising really as the Oliver Corporation was acquired by The White Motor Corporation of Oak Brook, Illinios, USA in 1960, and the name of Oliver relegated into history.

Accordingly, a preamble into the Oliver background will add significance and emphasis to the extraordinary resolution made by Ben Chifley in 1948, particularly considering the desperate shortage of tractors in Australia at that time.

James Oliver was born in Roxburghshire, Scotland, the son of a shepherd, in 1823. The family emigrated to the USA in 1835 and in 1855 James Oliver commenced manufacturing ploughs in premises he acquired at South Bend, Indiana. He could not at that time have envisaged the amazing fortunes that were to follow.

Joseph Oliver eventually took over control of the business, following his father's death, and by 1912 The Oliver Chilled Plow Works embraced an area of 60 acres, most of which was under roof! In 1929 Joseph was successful in achieving a remarkable amalgamation with several other thriving farm machinery manufacturers including Hart-Parr Tractor Co. and Nichols and Sheppard, both long established producers of tractors. The new conglomerate was to be known as The Oliver Corporation. In 1944 it acquired The Cleveland Tractor Company, manufacturers of an extensive line of Cletrac crawler tractors.

The Oliver Corporation was now one of the word's largest producers of farm machinery and tractors. Significant numbers of Oliver, Hart-Parr and Cletrac tractors had been sold into Australia since before the 1920's, earning a reputation for reliability and longevity.

Which brings us now to the Australian political involvement.

## THE MISSED OPPORTUNITY (OR 'YES PRIME MINISTER').

On the *30th June 1948,* Prime Minister Chifley received a letter from a Mr Leslie Dyke, who gave as his address 8a Castlereagh Street, Sydney. Mr. Dyke was a representative of The Oliver Corporation and had recently returned from discussions with the Oliver executives, the subject of which could have far reaching beneficial implications for Australia.

Mr Dyke requested a meeting with the Prime Minister for the purpose of outlining his company's aspirations to establish a tractor manufacturing and assembly plant in Australia. There was no suggestion of a request for federal funding or favours to assist with the project. However Mr Dyke did suggest it could be helpful if the Minister for Customs, Senator Courtice, could also be in attendance.

A reply was forthcoming on the *7th July* from the Acting Prime Minister, H. V. Evatt, who stated that the matter had been passed on to the Comptroller-General of Trade and Customs. There was no other comment.

As no further contact from the government was received by Mr Dyke, he again wrote to Prime Minister Chifley on the *7th September.* Obviously the Oliver executives were anxious

*The 1953 Oliver Fleetline Diesel 88, restored by the author, was a truly outstanding tractor. Its 45.15 h.p. 6 cyl. diesel engine was simply a masterpiece. Not only was it practically vibrationless at any r.p.m. and appeared to have a greater abundance of torque than its capacity would suggest, but it even looked good ! Careful consideration had been given to the layout of componentry including fuel lines and if ever an engine could be described as 'beautiful' then this was it.*

to obtain some response to their offer to invest millions in Australia and by so doing contribute considerably to the easing of the critical tractor shortage. The writer offered to meet with the Prime Minister either in Canberra or Sydney at a time convenient to the Prime Minister.

A letter dated *20th September* was forwarded from a Mr W. T. Turner of the Dept of Trade, Canberra, to a Mr J Garrett, Private Secretary to Prime Minister Chifley, advising that the matter of the Oliver proposal had been passed on to the Administrative Officer of the Central Import Licensing Branch.

Around ten weeks elapsed from the date of his original letter, before Mr Dyke finally received a blunt note from the Prime Minister on the *21st September*, stating that the Department of Trade and Customs was awaiting further information which had been requested from him. Infact, Mr Dyke had not been approached for further information! Nor was there any suggestion in the note of enthusiasm or appreciation for the Oliver proposal.

On the *25th January 1949*, the Prime Minister wrote to Mr. Dyke stating that a Mr Meere, the chairman of the Inter Dollar Committee, was of the opinion that the Oliver proposal should be further considered. The letter went on (and I quote) *In view of these developments I do not think anything would be gained by having a discussion with me. Your proposals will receive detailed consideration by the appropriate Commonwealth authorities when the further information requested has been furnished.*

From there the matter seemed to simply run out of steam.

## REFLECTIONS.

Several thoughts come to mind relative to the foregoing.

Firstly, I cannot envisage a more classic example of bureaucratic paper shuffling!

One would think that in 1948 the Commonwealth Government would have extended the welcome mat to a large corporation wishing to establish a tractor plant in Australia, particularly with regard to the acute shortage of tractors. The cold shoulder bureaucratic treatment I feel was astonishing – but probably typical of the impersonal treatment dished out by some bureaucrats.

The Oliver team must have been perplexed and possibly disenchanted by the Australian Government's apparent lack of interest in the proposal.

It should be remembered that the Oliver offer did not involve the acquisition of any Australian company or a request for government subsidies.

Would a similar situation be different today? Let's hope so – but I wonder if politics have really changed that much from the 1940's?

For the record, in the unlikely event I am challenged about

the accuracy of the Oliver saga, it so happens that copies of the documents are carefully preserved in my archives.

# FLOGGING AN OLD HORSE!!

Yes, I am fully aware that this epistle will disappoint many of my hitherto fans, who possibly thought I wasn't a bad sort of a bloke.  To them, my image may well be irreversibly tarnished. How could he write such nonsense, I hear them complaining! He must be past it!

Well, they may be right.  On the other hand maybe, just maybe, there will be those who will applaud my unyielding views.

So what am I raving on about?

Vintage tractor pulls – that's what!

And therein lies my problem. The vast majority of folk who flock to vintage tractor pulls think they are great!  Logically therefore they will be unhappy with my  opposing convictions.

I harbour a warm regard for the wonderful old idiosyncratic tractors that served us so well in days long gone.  Accordingly I cringe when I see them abused at competitive tractor pulls. Frequently I observe them driven in a manner that, when they were brand new, would have had an operator sacked on the spot and ordered off the property by a rightfully outraged farmer. And here they are in their Autumn years being flogged mercilessly, simply to provide entertainment for the uninitiated and a perverse kudos to their uncaring hoon drivers. (A bit over the top? Probably. But I am in one of my *Grumpy Old Men moods*).

Every weekend, legions of tractor enthusiasts joyfully head of to a vintage tractor rally, of which there are numerous all

*The Chamberlain in the picture has received an engine transplant, in the form of a 6 cylinder 100 h.p. Ford diesel. In its endeavour to pull 'The Old Poppa Stoppa' at a competitive tractor pull staged by the Sale Vintage Tractor Club, it is belching about as much smoke as one would expect from a coal fired vintage traction engine.*

around Australia. An increasingly popular segment of these events is the vintage tractor pulls.  For those who have not witnessed one of these unfortunate episodes, permit me to set the scene.

Old tired tractors, of all shapes and sizes, are dragged from their places of retirement and prepared for battle.

Firstly, pressure in tyres is reduced to an iniquitous 7 or 8 p.s.i., just enough to prevent the tyres from slipping on the rims but providing a ludicrous near flat profile in order to obtain maximum adhesion.

Next, an assortment of cast iron counterweights are attached to the rear wheels and/or axles.  There is a limit to how many can be physically fitted, so I know of some owners who deviously replace the sheet metal of the cockpit floor with three inch cast plate.   The luckless tractor will likely weigh twice its original weight, the significance being - the extra ballast will provide additional traction by reducing or eliminating wheel slip.

Now it is the engine's turn.  A tractor's petrol/kero engine, designed to run smoothly all day at a maximum eighteen hundred leisurely revs per minute, has its governor doctored and now screams at well over two thousand revs.  Larger jets are fitted to the carburettor to aid this punishing result.

Old diesel powered tractors have their engines similarly abused.  Fuel pumps are indelicately metamorphosed so that indecent amounts of diesel are squirted into each cylinder via transfigured injectors.

The vintage Lanz Bulldogs, with their single cylinder two-stroke crude- oil burning engines, are particular favourites among the tractor pull fraternity. It is a simple task to increase the fuel volume from the pump to the atomiser and fiddle with the governor, the result being clouds of putridity accompanied by flames, belching forth from their chimneys. Spectacular in the extreme for the spectators, but with the likely result of causing acid rain to fall upon New Zealand and a buoyant market for the manufacturers of hearing aids for deaf tractor drivers!

Upon arriving at a rally site, these hapless tractors await their turn to be hitched to a specially constructed mobile torture machine, in the form of a weighted sledge with wheels at the rear and skids at the front. The idea is for the tractor to drag the sledge along a track, usually of around seventy metres in length. As the tractor proceeds, the concrete weights on the sledge progressively move from the rear to the front of the contraption, thus creating an escalating load (or drawbar pull) requirement, increasing as the tractor advances along the course.

The tractors are separated into horse power classifications. Those which can haul the sledge the furthest distance, take out the honours in their class. But even here there can be surreptitious trickery.  The ratings are supposed to be based on *brake* horse power, yet frequently the lesser *drawbar* horse powers are submitted by an owner in an endeavour to have his tractor rated in a lower class.   If the scrutinisers are not awake, this gives him an unfair advantage in that class.

In addition to the crowd appeal of the single cylinder Bulldogs and Field Marshalls, a Chamberlain Super 90 with its ear piercing 2 stroke GM 371 diesel engine always attracts a hearty response.  A Super 90, even without any doctoring, is seldom outperformed.  Other vintage muscle machines include the Fiat 80R, the Oliver Super 99 and the Marshall MP6 with its 6 cylinder Leyland diesel.

There is no doubt, the crowds delight in these spectacles. Lots of smoke and howling engines, with tractors being brought to their knees or rearing dangerously, as they are flogged without mercy by their red necked operators.   But the spectators cheer, as did the Romans in the Coliseum when slaves were thrown to the lions.

Further, there is a definite danger factor associated with these tractor pulls.  For example, safety officers are supposed to observe the proceedings diligently and immediately stop a tractor that lifts its front wheels off the track, which is a prelude to rearing with the possibility of tipping over backwards.   Sadly, I have not infrequently observed safety officers yarning away with each other instead of focusing on the performance of a tractor.   A serious accident involving a tractor during a tractor pull would have far reaching consequences, particularly in the indeterminate area of insurance.

In addition, I am appalled when, at some rallies, spectators are permitted to line up within a few metres of the tractor pull course in order to shout their encouragements and obtain a close-up view.  Anyone who has experienced a tractor tyre blowing out or (worse) an overtaxed engine disintegrating, must share my very real concern about the lack of safety and understanding exhibited by the safety officers, for permitting such close proximity of the crowd.

v my chest, and despite everything, I am privileged to have umpteen bushy tailed friends throughout Australia who are devotee tractor pullers.   Hopefully this friendship is not now in the past tense.   Come on fellas.  I wasn't being personal.  If pressed I would even confess that not all tractor pullers are inconsiderate of their machines.  Putting a reasonable load on an unadulterated faithful old tractor for a short time – is ok, well sort of!

But I remain one of the minority who find tractor pulls pointless and without relevance to the true vintage tractor enthusiast.  It is often pointed out to me that tractors are only heaps of nuts and bolts and therefore have no feeling – so why not stress them to their maximum?  I suppose I cannot argue that logic.

But heck, to me an old tractor is like an old horse.  It should be treated with respect and indeed gratitude. Never flog an old horse!

*This single cylinder 45 h.p. KL Bulldog is hooked to a 300 h.p. Mack Truck, much to the delight of the crowd at the Rusty Iron Rally, held at the Wauchope Showground.  Not so happy was the driver of the Mack, when  the Bulldog reigned supreme in the pulling competition.*

# CABBAGE, URSUS AND MORE CABBAGE!

## WHY CABBAGE?

Perhaps it was a mistake visiting Poland in August 2013.

Over the years Margery and I have driven around most of Eastern Europe, doing our research thing on old tractors, but never before in Poland. So why did we decide to visit Poland last August? Simple! To carry out my long neglected research into the evolution of the historically interesting indigenous Polish tractor – the Ursus.

So why might it have been a mistake to go there? *My profound dislike of cabbage!*

Permit me to explain. Back in the Dark Ages, my presence was inflicted upon a certain school in Edinburgh. In the dining hall I would line up bearing my plate and proceed past the long table, behind which stood a row of apron clad kitchen staff each with a cauldron of 'stuff' which they dolloped on to each boy's proffered plate. The 'stuff' might range from watered-down yesterday's mince, gravy with floating blobs of meat which they somewhat euphemistically referred to as stew, greasy fried herring, or if we were fortunate – good old Scottish haggis. Plus there was an assortment of over-cooked vegetables.

But always at the end of the table was the dreaded cabbage woman, noted for her mop of untidy straggly grey hair. As she leaned over her steaming pot of cabbage mush, she repeatedly scratched her head, the action of which encouraged showers of dandruff and floating hairs to descend into her nauseating brew!

Can I be blamed for harbouring a lifelong detestation of cabbage?

I was therefore understandably dismayed when I discovered the national dish in Poland is cabbage. It comes in many forms. There is boiled cabbage, fried cabbage, roasted cabbage, pickled cabbage, cabbage broth, green cabbage, red cabbage, blue cabbage, cabbage omelette, cabbage dumplings, cabbage stew and even (would you believe) cabbage ice cream – to mention only a few!

To confuse matters, very few of the menus in Polish restaurants include English translations. Also, apart from in the glitzy hotels, most of the waitresses and waiters (woops – I am supposed to say 'wait persons') don't understand a word of English apart from 'hello' and 'you pay now'. And frankly the

Polish language is so complicated that I marvel the Poles can understand it themselves!

Not surprisingly I shed several kilos whilst in Poland.

## A BIG SURPRISE.

We picked up the rental car in central Berlin and navigated our way south for around twenty kilometres until joining the A12 autobahn, then headed east towards the Polish border.

Driving on the brilliant German autobahns is always a pleasure. The Germans in my opinion are the world's most skilled and considerate drivers. This is evidenced on the autobahns, most of which impose no speed limits – apart from trucks. The majority of cars hurtle along at speeds ranging from 120 k.p.h up to around 160 k.p.h, driven with a high

*A view through the car window of a stretch of the Polish super highways that are being constructed throughout the land.*

degree of safety and courtesy.  Many are driven much faster.

As we approached Poland, I alerted Margery to the fact that we should prepare for some fairly indifferent roads compared to those in Germany.  I had in mind our travels in such places as Serbia, Romania and Bulgaria.  How wrong I was!

The instant we crossed into Poland we encountered the most magnificent motorway we had experienced anywhere.  Better than the German autobahns, the Italian autostradas and even the interstates in the USA.  We were on the recently constructed  E30 which carved its way through strikingly beautiful undulating forest country, in the direction of distant Warsaw.  The road surface, the width of the lanes and the pristine landscaping has to be seen to be appreciated.

We were to learn that these super highways are being built all over the country. However our desire to experience at first hand the diverse farming communities, saw us also meandering along hundreds of kilometres of often pot holed or gravel roads connecting ancient villages and groups of farms.  In particular we were hoping to track down some early examples of Ursus tractors.

## URSUS.

The origins of Ursus tractors can be traced back to 1893.  In that year a group of seven Polish businessmen identified Russia as a potential market for the sale of stationary engines and ancillary equipment. Accordingly they established the Ursus factory in the Warsaw suburb of Czechowice and production commenced.

Quite remarkably and possibly without precedent, the initial capital of the company was identified as being the dowry of seven maidens who were the daughters of the seven founders! (I have confirmed the accuracy of this astonishing fact). Indeed the firm's trademark was the symbol 'P7P' which represented Posag 7 Panien i.e. *The Dowry of Seven Maidens.*

By 1913 a total of 6,000 engines of up to 450 h.p. had been supplied to the Russian Tsar. In 1918 the first prototype tractor was produced. Few records remain, but it appears to have been a copy, perhaps with licensing approval, of an International Titan 10-20. By 1927 a mere one hundred of these tractors had been manufactured.

In 1930 the company, now entitled Zaklady Mechaniczne Ursus, Inc., experienced unsolvable financial problems and was acquired by The State Engineering Factory, the manufacturer of military vehicles and weapons. By 1939, under its new ownership, Ursus had produced 737 tanks, 700 military tractors, and over 1,000 Ursus and Sauer cars and commercial vehicles, in addition to 2,500 motor cycles and an unknown quantity of aero engines.

During World War 2 the Ursus factory was destroyed by the rampaging Nazis, along with nearly every other building in Warsaw.  No other city was so utterly and needlessly gutted and ravaged during the war as Warsaw.  (Man's inhumanity to man)!

In 1945 and now under the harsh Soviet regime, the Ursus factory was rebuilt from scratch.  Agricultural tractors were urgently required. It was decided to manufacture copies of the German Lanz Bulldog Model P.  The 45 h.p. single cylinder crude oil- burning 2 stroke engined machine was designated the Ursus LB 45 and later the C 45.  Between 1947 and 1959, 60,000 of these rugged and dependable tractors were produced.

The first all Polish designed Ursus was the C325 released in 1957.  This was a well engineered lightweight, powered by a 2

*Pictured is a 1960s Ursus C 360 powered by a 4 cylinder diesel of Ursus own make.  Testimony to their excellence of design, hundreds are still daily working on Polish farms.*

cylinder upright diesel engine producing 25 h.p. On 29th June 1961, a C325 was submitted to The Nebraska Testing Facility in the USA. It performed admirably and returned an impressive drawbar pull of 1,743 lbs. at 3.89 m.p.h. Other diesel models soon followed. Although unremarkable in design, they proved ideal for the Polish farmers at that time. They were easy to maintain and utterly reliable.

In 1969 Ursus entered into an exchange of technology alliance with the Czechoslovakian Zetor tractor manufacturer and produced the 76 h.p. Ursus C385, which was virtually identical to the Zetor 8011.

A momentous agreement was signed with the giant Canadian Massey Ferguson group in 1974, enabling Ursus to manufacture MF tractors and Perkins diesel engines.

Fast forward to 1990. It was around this period that Massey Ferguson Australia Ltd. commenced importing Ursus tractors for the local Australian market. The Banner Lane MF factory in Coventry, UK, could not maintain an adequate supply chain and therefore the Polish manufactured units plugged the gap. Five models were offered ranging from 38 to 119 DIN h.p. However by far the biggest sellers were the 47 h.p. 3512/4 and the 60 h.p. 4512/4.

Sadly, during the 1990 decade sales of Ursus tractors declined rapidly. The company had acquired an unsurmountable amount of debt during its expansion programme of the 1980s. In 1996 it was unable to meet repayment commitments of 550 million zloties and over 700 creditors were written off! In 2007 the Turkish Uzel Holdings Group, who were AGCO Massey Ferguson licensees, agreed to acquire Ursus, then failed to proceed with the arrangement. Effectively this was the demise of a once proud and great Polish tractor organisation.

## CABBAGES AGAIN.

We spent nearly a month in Poland and warmed to the country and its people, despite their craving for cabbage and their enigmatical language. My linguistic skills could only extend to *Dozobaczenia* – I'll see you later – and – *Dzien dobra* – Good day.

Whilst most of our roaming was in rural Poland, we also spent some days in cities such as Poznan, Krakow and of course Warsaw. Although having a comprehensive historical knowledge of the horrendous victimisation the Nazis perpetrated upon the citizens of Warsaw, I was not prepared for the personal sobering psychological effect resulting from our visits to various war museums and shrines. The experience will remain with me forever.

So it was good to escape to the countryside again. Seemingly everywhere we wandered we came across ageing Ursus tractors working merrily in the fields.

Farming activity in Poland is heterogeneous in the extreme. The old Ursus tractors were mainly employed on small meticulously maintained arable farms, with obviously high yielding crops and rich tilthy soil. In addition to the old tractors, teams of labourers were frequently observed manually hoeing weeds, which reminded me of my youth in Scotland.

However, there was no escape for me! I estimate at least eighty per cent of the crops were (you guessed it) cabbage! Now I don't actually dislike cabbages when they are growing in a field. It is when they appear on a plate in front of me that I feel decidedly queasy.

Poland can also boast thousands of hectares of broad acre farming. Magnificent crops of grain are spread across the softly undulating landscape with high tech tractors including International, John Deere, Fendt, Massey Ferguson etc., busily doing their thing.

Our final days in Poland were spent in the south of the country in The High Tatra Mountains – no agriculture here! Then south west into the land of castles - beautiful Slovakia, the home of Zetor tractors and the historic capital  Bratislava graced by the gently flowing Danube meandering by.

# THE GOOD GEAR.

Seven months after placing the order, I recently took delivery of a diesel fuelled European car equipped with one of these new high tech. twin clutch semi-automatic transmissions. I have to admit, the brochure was right! The seamless gear changes can often only be detected by a close scrutiny of the rev counter.

Mind you, I have been subjected to ridicule from a handful of my alleged car club mates, who persist with the belief that a real driver prefers to change gears manually. Yet put them in a 1940s Austin truck, or my old 1928 Talbot, or better still – a 1920s Fordson Model F, with their soul destroying crash gearboxes and my car club compatriots would likely be in a state of extreme confusion. But of course, driving these vehicles with their idiosyncratic transmissions presents no problems for we ageing tractor folk!

Frankly, in these modern times I am surprised that anyone desires to own a family sedan with a manual-change gear box. After all, automatic transmissions have come a long way since the chewing gum autos of the 1960s.

So having vented my views relating to car transmissions, predictively my thoughts now turn to tractors and the 20 forward and 12 reverse gears that I originally thought was an overkill in my Landini. But guess what – I use every one of them!

However I can assure you, there were some wacky transmission designs in many of the early tractors.

## IN THE BEGINNING.

Apart from a few experimental machines, internal combustion engined tractors commenced their irrevocable march across the rural landscape during the first decade of the Twentieth Century. In the main they were obstreperous, clattering and often dangerous contraptions of mammoth proportions. The philosophy inherited from their steam powered cousins prevailed. In other words, it was believed the new generation oil powered tractors had to equate the weight and size of steam traction engines in order to be capable of pulling the broadacre implements of the era.

For example, a tractor weighing around 10 tons with a huge but inefficient engine (by today's standards) producing a mere 20 h.p., required at least half its power to simply propel the unit *without* a load.

The transmissions of these first tractors were crude in the extreme. Rough cast-iron gears were exposed to mud and grit, with the resultant rapid deterioration one would expect. Few manufacturers considered the frivolous idea of encasing the gears and having them running in oil!

Initially most tractors were only offered with one forward and one reverse gear, providing a plough speed of around 2 m.p.h. Eventually two forward speeds became the rule.

Three examples of the more idiosyncratic early tractor transmissions are examined hereunder.

## INTERNATIONAL FRICTION DRIVE.

International Harvester's first tractor was based on a design patented in 1903 by an engineer named Morton, who had established a factory in Chicago specialising in the design of tractor chassis and transmissions, to which customers could add an engine of their choice.

The International marketing team saw this as a rapid way of entering the tractor business, thus bypassing much of the design development. All that was necessary was to adapt their well accepted single cylinder open crankshaft 15 h.p. Famous engine to the Morton chassis. The massive engine featured an 8 inch bore and 14 inch stroke and developed its 15 h.p. at 240 r.p.m., the speed being regulated by a hit-and-miss spark governor.

But what made the big tractor technically fascinating was the method of delivering the power from the engine to both the forward and reverse drives. The operator was required to manipulate two large levers positioned by his left hand, the first of which actually moved the engine either forward or backward along the chassis, thus engaging the flywheel by friction to the forward drive. The second lever smoothly engaged the reverse drive.

Surprisingly the Friction Drive was a pleasure to operate, requiring very little physical effort compared to conventional power trains utilising the heavy clutch designs of the period. The action could be likened to that of a modern hydraulic shuttle control, such was the ease and smoothness of the operation.

## RONALDSON TIPPETT SUPER DRIVE.

The Victorian Ballarat firm of Ronaldson Bros. & Tippett introduced their Super Drive 18-30 in 1924. The design was based on the Illinois 18-30 produced in America by The Illinois Silo and Tractor Company of Bloomington. The unit was powered by a 30 h.p. Wisconsin engine, featuring 4 cylinders in 2 banks of 2.

Operating in Australia's torrid summer weather the tractor immediately encountered over heating difficulties. The problem was overcome by increasing the capacity of the radiator. The modified tractors were instantly recognisable by the profile of the heightened radiator header tank.

Interestingly, Australian engineers designed a special manifold which enabled the engine to be run on crude oil fuel, following the initial warm up with petrol. This added considerably to the appeal of the tractor, as crude oil was cheaper than petrol and even kerosene.

However undoubtedly the most interesting feature of the Super Drive was its 96 speed gearbox! Well – potentially 96 gears.

You see ostensibly the gear box provided 2 forward and 1 reverse speeds. *However*, this could be augmented by first removing a cover on the left side of the transmission and swapping the position of two 'pick-off' gears. This now offered the choice of 4 forward and 2 reverse speeds.

*But that is not all.* Ronaldson Tippett could supply an extensive range of 'pick-off' gears which, according to the operator's manual, could provide no less than 48 speeds under

*The author is at the controls of this Ronaldson Tippett 18-30 Super Drive, one of the many vintage tractors on display at the Pioneer Settlement Village, Swan Hill, Victoria. Remarkably an option of 96 forward gears are available. (Photo M. Daw)*

6.6 m.p.h. (The mind boggles!)

*But wait – there is more.* A choice of 2 rear wheel diameters was also available. So if one does the maths, there were theoretically 96 forward and 48 rear speeds. Undoubtedly a world record for *any* type of transportation.

The control of the clutch was also interesting. As the clutch was engaged, the engine revolutions were automatically increased. This was supposed to overcome the problem of the engine stalling when the clutch picked up the load. The downside was, the operator who was perched on a narrow wooden platform at the rear of the machine, had to really hang on as the machine jerked forward or be in serious risk of being jolted off, right in the path of the plough!

Fortunately most of the Super Drives were sold to grain farmers whose agricultural land was relatively flat and accordingly seldom had to apply the horrendous braking system.

If a gear was *disengaged* the brake would not operate! Think about it. Imagine a Super Drive negotiating a hill, either up or down, and it became necessary to change down to the lower gear. The perilous procedure would be to firstly apply the foot brake (there was no hand brake) with the hand clutch disengaged but the gear still engaged. Then, presumably with the aid of an offsider, a block would be placed against the wheel, preventing the tractor from taking off whilst the operator moved the gear shift through neutral and into the other gear.  Scary stuff!

## FOWLER REIN DRIVE.

In the 1920s the horse still reigned supreme on Australian farms.  Even the bargain basement price of a Fordson was not a sufficient inducement for the average farmer to retire his team to back paddock and invest in one of these new fangled intimidating tractors.  I mean to say, one would have to comprehend the incomprehensible gearbox-clutch routine whilst wrestling with a steering wheel!  And then there were all these leavers and doodahs to worry about!

On the other hand, it is a well known fact that the only controls necessary for the horse is a pair of reins.  Such was the reasoning of the many farmers who were nervous of tractors.

Then up sprung Cornelius Murname, a design engineer from Melbourne.  He had a weird but clever idea of how to overcome this farmer resistance to the complications of driving a tractor.  He presented his patents to John Fowler and Company of Leeds, England, who were in the business of manufacturing very large heavy weight tractors and thus were losing business to the multitudes of lighter machines that were becoming increasingly popular.

In 1924 the Fowler Rein Drive was unveiled at The Royal Agricultural Show at Leicester, where it won a gold medal.

However that was its one and only day of fame. It turned out that in fact farmers were not enamoured by the prospect of driving a tractor solely by means of a pair of reins, because whilst the driver was obliged to be perched on a sort of trailing buggy, or astride a trailed implement, a rope rein was the only contact he had with the tractor. Different tugs on either the left or right rein controlled the clutch, gearbox and steering.

Considering the overall dimensions of the rig, it was powered by an engine of somewhat alarming proportions. It was a liquid cooled V twin configuration with a 5.75 x 7.5 inch bore and stroke and developed 32 h.p. at 1,000 r.p.m. Designed by Fowler, it was virtually half an engine that was originally deployed for propelling a military tank.

The accompanying illustration is of possibly the sole remaining Rein Drive and is on public display at the Pioneer Park Museum at Parkes, NSW. This rare exhibit was restored by Stewart Nash and is but one of scores of magnificent tractor artefacts lovingly cared for by a team of dedicated enthusiasts.

A few years ago I was offered a drive of the Parkes Rein Drive. Despite the fact that I have been privileged to have operated untold numbers of weird old tractors around the world, this was the scariest I have ever encountered. I mean – a pair of reins!!! What if the rope broke? It is no small wonder that the Fowler Rein Drive was a commercial flop.

# TRACTOR LOADER-BACKHOES REMEMBERED.

The winter of 1961  was the year I was appointed as a young 26 year old sales manager in the employment of Lough Equipment Pty. Ltd. of Artarmon, on Sydney's North Shore.

I was probably far too immature to occupy such a lofty position, with the necessary degree of dignity and professionalism. But the truth is, Eric Lough thought I was God's gift to the world of earthmoving equipment!

You see there had been a week long construction equipment trade show held mid-winter at The Sydney Showground. The event extended into the evenings in order that busy contractors and civil engineers could attend after work.  The Thursday evening happened to be bitterly cold, with strong winds and frequent rain storms lashing the Showground.

Accordingly, sales folk and the paying public all made a beeline for the comparative warmth and shelter of the pavilions – much to the satisfaction of the indoor exhibitors.

The outside exhibits remained unattended.  That is, with the exception of The Cumberland Tractors Pty. Ltd. stand.

As a lowly industrial representative, employed by Cumberland Tractors, selling the recently introduced Massey Ferguson range of earthmoving machinery, I had been rostered for the Thursday evening shift.  Frankly, this suited me fine, as I was able to ensconce myself within the cosy confines of the cabin of an MF 11 loader, turn up the heat and read the latest edition of the somewhat ribald *Kings Cross Whisper*, left behind by an earlier visiting council engineer.

An umbrella approached, and thinking it could be Harry Clarke my sales manager, I guiltily scrambled out of the heated cabin and pretended that I had diligently been braving the rain all evening, in anticipation that a potential buyer might arrive on the scene.  It wasn't Harry.  It was the well known and popular earthmoving machinery identity Eric Lough, making his way from his car to his firm's stand in one of the pavilions.

"Good God Ian, what are you doing being out in this weather?"

"Oh hello Mr Lough" I responded, having to shout over the gale.  "Well I wouldn't like to miss a prospect, should one happen to brave the storm."

"Now that's what I call true dedication to the job!" he said, regarding me speculatively.  "I am very impressed" he added over his shoulder, as he hurried on.

An hour or so later, it was time to go home. I folded The Whisper and stiffly climbed out of the cabin. I was just about to lock up when once again the umbrella with the tall figure of Eric Lough striding along, came into view.

"Don't tell me you are still here" he exclaimed. "Has it been worth being frozen all night?"

"Well, with two orders under my belt, it certainly has been worth while" I replied jokingly.

It was a joke, right?

But I realised later, my attempt at witticism had miscued because it became apparent that Eric Lough had believed what I had said. Two days later I received a phone call from him, wondering if I would like to join Lough Equipment as Sales Manager!

I wrestled with my conscience, but for the record I *did* reveal the truth of the matter to him. But he didn't hesitate and remarked that he appreciated my honesty, and the offer stood. I enjoyed my work at Cumberland Tractors and my association with the Massey Ferguson range, but here was an invitation to leap ahead into a prestigious executive position, at a relatively young age. I accepted the offer.

(For the record – I returned to Cumberland Tractors a few years later in the roll of General Manager).

# THE WHITLOCK DINKUM DIGGER.

Lough Equipment held a variety of construction machinery agencies, including Thwaites dumpers, Hydor compressors, Greens rollers, Mitsubishi crawler loaders and Wylie asphalt plants. However the firm's main claim to fame was its Whitlock franchise.

Whitlock Bros. Ltd. of Great Yeldham, Kent, manufactured the range of Whitlock Dinkum Diggers. The Dinkum Diggers, a Scottish invention, were the world's first hydraulic backhhoes, originally designed for attachment to Ferguson and Fordson farm tractors. They were also the very first hydraulic backhoes available in Australia, when introduced by Eric Lough in the mid 1950s.

From a modern day perspective, they were a shocking piece of gear – a hydraulic nightmare and an engineering calamity. But despite this, they worked! Plumbers and construction contractors loved them, no doubt because the alternative was a pick and shovel!

A Mark 2 Dinkum Digger arrived on the scene in 1961, built locally under licence to Whitlock Bros Ltd., by Harry Walters & Co. of Auburn. Certainly it was less antiquated than the original, but there was a penalty to pay. It was grossly overweight. Particularly when bolted on to a poor little Ferguson 35! A front-end loader, if fitted, helped the situation by acting as a counter weight.

A plumber in Wagga Wagga purchased a Mark 2, much to his later regret. He elected not to fit a loader to the front end of the rig – a big mistake! On its first outing, a senior police sergeant was amazed when he observed a tractor proceeding along Fitzmaurice Street "...in a highly dangerous erratic manner with its front wheels bouncing two feet into the air". He assumed the operator was either intoxicated or a lunatic.

Majestically, this protector of the law, strode out into the path of the oncoming hazardous apparatus with his palm raised, in a manner that left no doubt that it was a command to stop.

With notebook at the ready he confronted the offending driver, only to discover with a shock, that it was his brother-in-law the plumber. The sergeant was accompanied by a constable, who of course was a witness to the whole affair, and although greatly embarrassed, the unfortunate sergeant had no alternative but to risk a family feud by booking his brother-in-law for dangerous driving!

Brother-in-law was not amused and successfully litigated against Lough Equipment for supplying an unroadworthy vehicle. This resulted in Eric Lough, his son Richard, who was the company accountant, an expensive Sydney barrister, and myself, cooling our heels in the lounge room of Romano's Hotel opposite the Wagga Wagga Court House, for two long days, waiting to be called as witnesses for the defence. (What possible defence?  But I guess the legal guy had to *seem* to be earning his keep).

Thankfully and wisely Eric Lough got cold feet and instructed the barrister to seek an out-of-court settlement. In the late afternoon we boarded the East West Airline DC3 for Sydney, having agreed to refund the cost of the Dinkum Digger plus the slightly placated plumber's legal expenses.

## THE BIG WHITLOCKS.

The larger of the Whitlock range were in fact big powerful backhoes built on special industrial Fordsons.  They were never outperformed when asked to compete against their Australian built competitors, such as Cranvel, Ace, and Steelweld, but they were not without problems.

Because of the Whitlock trapezium design the main boom, jib, etc., incorporated a bewildering number of pins and bushes.  Each of these required frequent and liberal squirts from a yukky grease gun.  Further, the bushes seemed to be manufactured from an absurdly soft bronze material, possibly sourced from Outer Mongolia, guaranteed to wear out as you watched!  Yet they were priced as if they were made of gold.

Also, the Whitlock slew mechanism utilised steel cables. At the end of the 20 foot boom, the big 36 inch bucket heaped with wet clay, weighed in at around a ton.  Imagine the scenario on a confined work site with the heaped bucket slewing in an arc at full flight – and the cable snaps! Not a happy thought.

A further expensive complication for unsuspecting owners resulted from the weight of the big Whitlocks, particularly the 7 ton Model 60A. The tractor axle housing flanges were simply not up to the job and had a tedious habit of cracking. This of course led to yet another loss of income earning down-time.

However, it should be emphasised that the company CEO, Eric Lough, was a gentleman of considerable integrity and stood behind every item of equipment sold by his firm. If he considered it unjust, in situations where Whitlock failed to

*The flagship of the Whitlock range was the Whitlock 60A. Undoubtedly, until the introduction of the JCB 4C, the 60A was by far the biggest capacity tractor mounted backhoe available. But it too was an overly heavy implement for the mounting to, in this case, the industrialised Fordson Super Major with its matching Whitlock Frontend Loader. When water ballasted the rig weighed around 7 tonnes, a figure which far exceeded the structural limits of the Fordson. The resulting frequency of cracked rear axle housing flanges was indicative of the stresses placed upon this (basically) agricultural tractor. (IMJ archives)*

accept a problem as a warranty claim, he was sympathetic to the owner and invariably authorised a free repair, charged back to Lough Equipment.

It should also be stated that despite the above mentioned technical problems, the majority of owners loved their machines. The hourly production from a 60A far outclassed any other digger. It was not until the introduction of hydraulic excavators, that greater output figures could be achieved.

## COUNCIL DEMONSTRATIONS.

A considerable number of the big Whitlocks were purchased by local government councils. This involved submitting tenders, frequently followed by requests to submit the digger to a comparative demonstration. Providing our tendered price was reasonably competitive, Lough Equipment invariably landed the deal. As mentioned earlier, no other backhoe loader combination could outperform the big Whitlocks.

Very often these council demonstrations were a farce, particularly when Sydney area municipalities were concerned. The assorted aldermen (or should it be *alderpersons*?) saw these demos. as a form of entertainment and an excuse to have a lunch at the ratepayers' expense, following the proceedings.

Visualise the scene. Three or four backhoes digging away at a council work site and a gaggle of aldermen, very often including the local chemist, the fishmonger, the baker and the candlestick maker, none of whom had a clue about machinery and earthworks, all nodding knowingly and wishing it was all over so that they could settle down to a long lunch.

The irony was, they each had a vote on which machine should be purchased!  Fortunately for me, it was always patently obvious to even a simpleton (yes, even the aldermen) that the Whitlock was the obvious choice.  Despite this, on account of price, it did not follow that every council purchase was necessarily a Whitlock.  But Lough Equipment certainly enjoyed a healthy percentage of the total sales.

However, there was a new challenge on the horizon.  Its name was JCB!

## JCB.

Following their perusal by Eric Lough, the 1964 editions of the British construction equipment magazines routinely found their way onto my cluttered desk.  For months we had been noting the dramatic escalation throughout the UK and Europe of the promotion and resultant  rampant marketing penetration of JCB loader backhoes.

We also noted that JCB had embarked on a global expansion of establishing importers in North America and  the Asia Pacific region.  Rather ominously we also detected that Whitlock sales nose dived wherever JCB established a toehold.

So who and what was behind this JCB phenomenon?

In 1945, Joseph Cyril Bamford constructed a farm trailer in his tiny lockup garage, using steel sheeting that had been part of an air raid shelter, which he welded together with a second hand 50 shilling welder! He presented it on market day in his Staffordshire village of Uttoxeter and sold it to a local farmer for the princely sum of £90. Over the next few hectic years of manufacturing, firstly in a coal yard and then in an abandoned cheese factory, his range expanded to front end loaders fitted mainly to Fordson tractors. In 1953 the first JCB loader backhoe was unveiled.

By 1964 JCB backhoes were the world's number one in the sales charts! A phenomenal achievement unparalleled in the construction machinery industry. Joseph Cyril Bamford was a truly remarkable entrepreneur!

## A NEW DIRECTION.

Back at Artarmon, Eric Lough was rightfully concerned that one day soon, JCB loader backhoes would appear in Australia and this could decimate our Whitlock sales. Therefore with considerable vision and (dare I say) cunning, he surreptitiously put out feelers to JCB in England.

In point of fact, we had been becoming increasingly disenchanted with our Whitlock franchise. Eric Tittley, the Australian Whitlock factory representative had his office at the remote rural hamlet of Dooralong, of all places! In 1964 the Dooralong telephone exchange only functioned a few hours each day. Even in the dedicated 'open' hours, if the exchange lady happened to be out feeding the chooks or had fallen asleep whilst watching a repeat of *Dragnet*, then Mr Tittley was off the radar.

Not a happy situation, as all communications with the factory had to be channelled through him. Accordingly, being unable to contact Mr Tittley could prove extremely frustrating, particularly if he was urgently required by Gib Gospal, the Lough service manager, to send a cable to the UK to chase up a rapidly mushrooming log of warranty claims, or to locate a missing consignment of desperately needed spare parts.

Harry Wilson was the JCB Marketing Director. In response to a letter from Eric Lough, he had cabled ahead and set up an appointment to meet us at Artarmon. Following a round table discussion, we knew he was our type of individual. We warmed to his ready smile and were impressed by his grasp of the construction equipment situation in Australia, even though this was his first visit to the Antipodes. His enthusiasm for the JCB product was persuasive and infectious. Further, he was not a prevaricator, but obviously a man of rapid decision.

I cannot recall the actual time frame involved, but in due course a franchise agreement was signed with JCB. Cartons of lavish brochures and promotional material began arriving.

These had to be hastily hidden away in the back room if Mr Tittley was due to arrive, as no announcements had been made regarding our new direction. Well, I mean to say, we still had a few Whitlock machines to quit. Although Mr Tittley started complaining about the lack of forward orders for new Whitlock stock. All's fair in love and war, they say.

Harry Wilson had investigated Lough Equipment thoroughly, he told us later and was impressed by our company's dynamism. So not only was he acquiring a switched on Australian distributor, but was also eliminating Whitlock as a sales opponent. Although the latter did not actually eventuate as Ken Coles Pty. Ltd., a backhoe hire organisation, picked up the Whitlock agency, which soldiered on for another *few* years.

## THE PRESS REPORT.

Eric Lough instructed me to prepare a press release, which was circulated in October 1965. It read as follows:

*Following 12 months market research and thorough investigation of excavator (backhoe) trends throughout Europe and America, the Directors of Lough Equipment Pty. Ltd. have announced that on 15th October 1965 their company will relinquish the Australian Whitlock franchise and will commence the distribution of JCB equipment.*

*It is with great reluctance that Lough Equipment finds it necessary to abandon the Whitlock association in favour of JCB,* as it is not possible to be associated with a product over the years, to foster its development and promote its sales, without creating an attachment towards it. Notwithstanding this, it has always been this company's policy to offer the very best equipment available to the Australian earthmoving industry, and in order to continue this earnest desire we have entered into mutually satisfactory arrangements with J.C. Bamford (Excavators) Ltd., the world's largest manufacturer of rubber tyred excavators.

A few weeks prior to the press release, Eric Tittley was advised of the state of affairs. Frankly, I felt sorry for him. He was actually a decent but naive gentleman of the old school. He was an accountant by training with no expertise in marketing. Seemingly the wrong man for the job, in this cut throat machinery business.

## JCB DAYS.

There were no less than five backhoes in the JCB range, all of which were equipped with front loaders, apart from the JCB 1, which had a front mounted back filling blade. All were fitted with lock up cabins and all were marketed by Lough Equipment in Australia.

The JCB 1 was an unorthodox lightweight machine, powered by a Petter 20 h.p. diesel.

The JCB 2B was aimed at the Massey Ferguson 220 market

and was mounted on a modified 44 h.p. Nuffield tractor. It had the amazing feature of being able to have the backhoe disconnected in two minutes! During a demonstration to the Sydney Metropolitan Water and Sewerage Board, one of the senior engineers failed to pay attention, and when he refocussed on what was going on, he was shocked to find the backhoe had 'broken off' from the tractor. The laugh was on him!

The JCB 3 was bigger capacity than the 2B but mounted on the same tractor, without the advantage of being able to be disconnected. It seemed to me to be a misfit in the JCB range. Others agreed and only a few were sold.

The JCB 3C was the jewel in the JCB stable. Mounted on an industrialised 60 h.p. Nuffield, it was a brilliantly capable and purposeful looking unit. Within two years of its launch in 1963 over 3000 had been sold.

The JCB 4C was a massive machine that could certainly pull out the dirt, but it was unwieldy and even dangerous to drive on the road. It was also too big to fit on a tip truck – a major disadvantage. The 60 h.p. Nuffield was hard pressed to support the bulk of the unit.

In 1965 JCB introduced its stylish and capable 360 degree slew, track type hydraulic excavator – the JCB 7, powered by a Ford 96 h.p. diesel engine. Upon its arrival in Australia, it was claimed to be the most technically advanced excavator of its type.  It represented a new experience for us at Loughs.  For example, getting one in and out of the workshop, which was below ground level, proved a real challenge.

The first JCB 7 sold, was purchased by Tipping Bros. of Chatswood. Carlotta Street was blocked for over an hour as we struggled to load it onto a *side* load low loader, much to the irritation of scores of vehicles that were obliged to back out onto The Pacific Highway.  Eventually it got underway but its height resulted in around twenty power wires being torn from their poles, all the way along Carlotta street.  The low loader driver was blissfully unaware of the mayhem he was creating and the tangle of wires that trailed behind his rig.

## FINALLY.

The Lough association with JCB was a happy one.  Joe Bamford sent his 21 year old son Anthony to Sydney for an acquaintance with the Australian earthmoving scene.

As was our custom with all visiting industry dignitaries, Margery and I invited him to our home for dinner.  He was accompanied by an old school friend, who was working his way round Australia and was currently employed as washer-up in a Kings Cross restaurant.  Imagine our surprise when we learnt he was The Honourable Jeremy Sykes, the youngest son of a prominent English peer!

Anthony proved to be a likeable young man with a keen

sense of humour. Margery originally came from Staffordshire, therefore she and Anthony got on famously and waxed on about villages and watering holes familiar to them both which, at that time, meant nothing to me.

Today, <u>Sir</u> Anthony Bamford is a multi billionaire and Chairman of JCB since 1976, following the retirement of his father Joseph Cyril, who died in 2001. Quite remarkably JCB remains a family owned business, with 18 factories spread throughout the UK, Brazil, Germany, North and South America, China and India. A range of 250 products are produced by a work force of around 7,000 and sold into 150 countries.

JCB regularly sponsors numerous charitable and research activities, plus vintage car race events, around the world. In 2010 The JCB Academy, a new secondary school located in Rocester, UK, welcomed its first pupils.

In 2006 the JCB Dieselmax, propelled by twin JCB diesel engines, broke the diesel engine world land speed record by achieving a staggering 529 k.m./h.

In conclusion, although my involvement was but a miniscule cog in the history of JCB, I remember fondly my brief association with possibly the most enterprising and successful construction equipment manufacturer the world has ever known. Yes, these were good days.

# THE ROAD TO THE ISLES.

## THE LURE OF THE HUNT.

Margery and I made one of our periodic trips to the land of my birth – Scotland, specifically on this occasion for the purpose of researching material for my latest tractor book. Whilst in Edinburgh I looked up Donald, an old school friend. Upon learning of my assignment he assured me that quite remarkably there was a magnificent tractor collection out on one of the islands.

Donald was a trifle vague as to *which* island, but he was *absolutely* certain it was one of the Outer Hebrides. Or at least he *thought* so! He assured me his informant was totally trustworthy. Being unable to resist the lure of a tractor treasure hunt, the very next day I pointed the Hertz Rover north along the M9, skirting Stirling and then northwest on the A84 and up through the glens of Crianlarich on our journey to The Road to The Isles.

It was late afternoon when we arrived at the West Coast town of Oban – the home port of the Caledonian MacBrayne island ferries. Following the signs, I navigated the car up a ramp and parked it as directed on the deck of a small weather beaten vessel. It departed Oban late evening and was scheduled to make a brief stop at Barra, a tiny rugged rock inhabited by a handful of stalwart citizens, at around 8 a.m. the next morning.

Barra is the most southerly of the chain of islands that make up The Outer Hebrides, which extend well out into the cold grey Atlantic Ocean. I did not anticipate for a moment that the alleged tractor collection would be located on this remote small island. A brief inquiry directed to the local harbour master confirmed this was so. Logically then it had to be further north in the chain of islands.

Following an exchange of mail and the handing over of newspapers, the sturdy little boat threaded its way past ominous fangs of forbidding rocks and departed North for South Uist, the next main island in the group, passing Eriskay (steeped in Bonnie Prince Charlie history) on its port bow. Solid no-nonsense wooden hulled fishing boats, with broad beams and deep drafts, hastened past in the opposite direction to some distant grounds, burying their high prows in the rising seas, then after an interminable delay it would seem, rearing up to attack the next assaulting wave.

# SOUTH UIST.

Lochboisdale is located on the Southern tip of South Uist and is largely just a short main street of grey houses which lead down to the stone jetty. The religion here is predominantly Roman Catholic, a legacy of the fact that many of the inhabitants are descendants from Spanish shipwrecked seamen, who narrowly escaped from the remnants of the stricken Spanish Armada. Swarthy dark eyed features are evidence of their ancestry. Gaelic is the commonly used language, but English is well understood, albeit sometimes reluctantly.

There was a fierce biting wind, accompanied by slanting sleet, sweeping in from the Atlantic as Margery and I entered the general store which also served as the Lochboisdale Post Office.

"You'll no be findin mony trectors on South Uist, so yill not" replied the wee wifey in response to my query. "If onybody wid ken it wid be Auld Wullie MacDuff up at Staoinebrig."

Staoinebrig, a group of stone houses and crofts, fronted a magnificent stretch of pristine windswept beach, onto which thundered giant waves that began their long journey from 4,000 miles west across the Atlantic. Auld Wullie, was not difficult to find. He was indeed old, but also hard of hearing and frowned when I addressed him in English. I repeated my question regarding a tractor collection, but this time reverting to the broad Fife dialect of my childhood, which I rightly imagined might help the situation.

He responded knowingly with his lilting island vernacular. "Ach, thurisnay nay trectir collection here aboot, so thur's no. Thur micht be ane up in Lewis. Ah hear tell thuv got lots o' trectors on the ferms there, so Ah do"

Great! Lewis is the most northerly island in the chain.

Within the warmth of the Rover we again headed north along the lonely narrow island road which wound its way through the bleak, flat and treeless landscape. We passed through Mingearraidh, Flora Macdonald's birth place. But the island extends only to Lochdar. The next island in the group is Benbecula connected to South Uist by a causeway across the sand, negotiable at low tide!

*A 'black hoose' (thatched cottage) on the isle of South Uist. Note the roped bricks securing the thatching in place.*

Benbecula (the name Robert Louis Stevenson borrowed for the infamous sailing ship in his novel Kidnapped) was reached at low tide without drama. However it was now approaching the gloaming (evening) and a place to rest-up for the night became a priority.

We were relieved therefore when we discovered a neat and tidy crofter's cottage along the road which advertised 'High Tea, Bed and Breakfast'. I imagined the high tea would be typically a Scottish feast, the sheets would be clean and the breakfast sustaining. I was proved correct in all three. But alas, our host knew nothing of any tractor collection on Benbecula! I was seriously beginning to doubt the integrity of Donald's informant.

Yet another low level causeway had to be negotiated in order to cross to the next island - North Uist.

## NORTH UIST.

The character of North Uist differed in an obscure way to that of South Uist and Benbecula. Interestingly, contrary to the South Uist predominance of Catholicism, the religion here was fiercely Protestant. But again Gaelic was the normal tongue. The island is nearly as broad as it is long – around ten kilometres from east to west, but embracing scores of inland and sea lochs.

The only village on North Uist is Lochmaddy. Like all the

Hebridean harbours, the approach from the sea is unbelievably hazardous. Only an experienced island skipper would be capable of navigating a vessel between the black rocks and dangerous shoals. Thirty foot tides, treacherous currents and frequent foaming seas add to the tribulations of an approaching skipper. It was from this place, the morning following our arrival, that I would be obliged to have the car loaded onto yet another ferry in order to move on to Harris, the next island where we would resume the search for the illusionary tractor collection.

In the meantime I booked a double room at The Lochmaddy Hotel, a fine traditional Scottish inn, possibly built during the era of the latter Stuart reigns. The amicable publican was apparently the font of all knowledge appertaining to this part of the Hebrides and shook his head when I raised the subject of a tractor collection.

"No, not aroond here" he replied. "But there is aye a wee bitty of a chance there could be some of thae tractors hidden away on a ferm up in Lewis But I dinnae think so" he added morosely. My confidence went into a rapid decline.

We dined that evening on a veritable banquet consisting of a gargantuan bowl of Scottish broth, followed by grilled lobster, then whisky embellished haggis with champed neeps and tatties (mashed turnip and potatoes), followed by oatcakes with Isle of Mull aged cheese. This of course was more than an adequate sufficiency, following the consumption of which, we retired to the lounge bar. There we were treated to a wee glass of Drambuie, that alluring Highland liqueur, courtesy of our loquacious publican.

He asked if we had noticed the cairn high up on what is the only mountain on North Uist, and located to the east of the hotel. No, we had not, possibly due to the mist surrounding the summit of the conical peak, which we learned was named Li a Tuath. He then recounted an amazing story.

Apparently a U.S. Airforce bomber, being ferried to Scotland during World War 2, slammed into the side of the mountain during a snow blizzard. The publican, a mere tousle haired youth at the time, was one of several volunteers to scale the slope in the remote hope of perhaps finding a survivor. Predictably, it was discovered that the entire crew had died in the crash. However some personal effects were salvaged and transported down to the village. Among these was the identity card of one of the American airmen.

The publican's eyes glazed over as he told us that by a staggering and tragic coincidence, the airman in question was identified as being the grandson o f a local sheep herder, who for years had acquired the rights to graze his flock upon the grass and lichen growing up the sides of that very mountain! The odds of such a coincidence are incalculable.

# HARRIS AND LEWIS.

With considerable difficulty, the next morning I drove the Rover up the ramp onto the heaving deck of the interconnecting island ferry. Upon inquiring the precise time of the departure for Harris, the well weathered elderly deckhand replied "Well that will depend, do you see?"

'Upon what?' I asked.

"Och the skipper is awa having a wee cup o' tea wi his dauchter who lives yonder" with a wave of his hand in the general direction of the village, "and when he is ready he'll wander doon and no doot have us cast off. Thurs nae hurry" he added.

How I envied this old man's wonderful philosophy. What a pleasing contrast to our life being ruled by the clock and every thing has to be at a rush!

Tarbert, the principal village of Harris, is positioned on a narrow neck of land between North and South Harris. The voyage took around two hours, no doubt slowed somewhat by the giant swells into which the game craft was subjected. The Rover was disembarked without incident and I drove it up through North Harris in the direction of Lewis. Here the terrain was quite different to that we had experienced hitherto on the islands. There were misty hills and bent trees through which the narrow road followed its way north.

It is interesting to contemplate that Harris is famous around the world for the cloth woven by the island folk, known as Harris Tweed. Some of the most expensive suits sold in London, New York and elsewhere are tailored from Harris Tweed!

Eventually we crossed the dividing line between North Harris and Lewis. Stornoway, the main centre and port of Lewis, was our destination and hopefully the place where we would finally learn the location of the tractor collection.

I confess to being diverted from our main purpose, when we took time out on our way to Stornoway to inspect a historic site of great antiquity. In order to reach our objective we were obliged to scramble precariously across a somewhat hair raising pedestrian rocky causeway, which extended out to the centre of a small black loch.

The purpose of this uncustomed alacrity was to explore a subterranean primeval fort which allegedly pre-dated the Egyptian pyramids! And there it crouched, I imagined precisely as it existed millenniums ago! Built entirely of massive granite boulders, it featured two metre thick walls and the cellar-like accommodation was protected above by a similar dimension of rock slabs. It gave the impression of being able to withstand a nuclear attack. A lonely spooky place indeed with the wheeling seagulls and the occasional splash of a salmon the only evidence of life.

Upon our arrival at Stornoway I parked the car outside a

handsome house to which were appended two signs. The one indicated that a part of the house was a veterinary surgery and the other that it was a high class bed and breakfast establishment. Excellent, for it would provide accommodation but also a vet would know all the crofters on the island and logically the location of a tractor collection. I was again feeling confident that my search was about to be justified.

"No Mr Johnston, I can assure you that if such a collection existed here, I would know of it, because I too have an interest in old tractors". These were the sobering words of the vet.

"Well I guess we may as well resign ourselves to returning to the Mainland tomorrow empty handed" I rejoined. "At least we have had a good tour of The Outer Hebrides" I acknowledged philosophically .

"I am sorry Mr Johnston. Tomorrow is The Sabbath and no ferries are permitted to enter or depart these waters on the Sabbath. Also I should alert you to the fact that all commercial activities cease trading – including restaurants, shops, even bed and breakfasts, such as ours, do not serve any meals whatsoever" he explained apologetically. "I advise you to purchase some food today to see you through."

He must have observed the look of dismay upon my countenance. After a pause he added conspiratorially "But if nobody knows, then I shall arrange to have a breakfast tray sitting outside your room tomorrow morning." I thanked him.

In actual fact we were half prepared for this strange custom, as we knew, that at the time of our visit, The Scottish Free Church (not to be confused with other Scottish Protestant religions) totally influenced the more-or-less entire community of Lewis. Its strict Puritanical edicts did not permit any commercial activities to occur on a Sunday. Further, all members of the Church were expected to spend the entire day in worship.

So on the Sunday we further explored Lewis, including driving to the most northerly point - The Butt of Lewis - where seals and sea otters displayed their incredible athleticism. Along the way we encountered small buses collecting families from their crofts to be transported to their places of worship.

On Monday morning, resigned and somewhat disconsolate at my failure to locate the tractor collection, I drove the car onto yet another Caledonian MacBrayne's island ferry, this time heading east across The Minch to the West Highland mainland and in particular the small fishing port of Ullapool, which squats at the entrance to Loch Broom. The voyage took around three hours. We booked into a harbour side hotel.

I was not aware, that the following day, I would have an unexpected encounter which would provide new clues of the likely location of the ambiguous collection - and yes it was indeed on a Scottish island, but far distant from The Outer Hebrides!

# HOT ON THE TRAIL.

The following morning we breakfasted on porridge, local kippers and oat cakes. Suitably nourished, I wandered down the jetty and spent an enjoyable half hour inspecting the colourful array of stout small craft secured by their hawsers to the sturdy bollards.

On returning to the hotel, I found Margery engaged in conversation with a middle aged chap I had noticed earlier in the dining room. I was introduced and learnt that Margery had been telling him of our failed mission to The Outer Hebrides.

"No, there are certainly no tractor collections out in The Hebrides" said the man emphatically.

"You speak with conviction" I said, observing him speculatively. "Have you an interest in old tractors?"

"As a matter of fact I have and happen to own a couple of restored old Fordsons" he responded. "But I am also a co-ordinator with The North of Scotland Regional Tourist Bureau and travel around extensively. It is my business to know what goes on in my territory" he added. Then continued, "The only collection you will find on an island up here in the North is a long way from Ullapool I'm afraid".

I took a deep breath. "Give me the bad news."

"Well I know there is a collection on one of the islands in the Orkney group, and as I said, that is a mighty long way from here" he replied apologetically. "But I also know that if you are searching for a tractor collection on a Scottish island, then I guarantee that is the only one.  But I am sorry I can't be more specific as to *which* Orkney island."

He spoke confidently and I had no reason to doubt the accuracy of his information. Obviously Donald had got his islands mixed up, which should not have surprised me as he was bit of a dunderhead at school!  But now the challenge of tracking down this allusive island tractor collection was becoming an obsession.  For all I knew it might consist of merely two common-or-garden Fergusons!  But I obstinately simply *had* to find out!

Margery contemplated for a moment, then with a twinkle in her eye said "No use standing around.  You had better get us to Orkney."

## THE DRIVE NORTH.

Ullapool lies at the head of Loch Broom in the shadow of the towering mass of Beinn Eiledeach.  The only road north is the A835 which steadily curves its way up through the Cromalt Hills to Ledmore, where it is joined by the road that leads from Inverness and Tain.  The Rover negotiated the breathtakingly beautiful glens and steep passes effortlessly.  The descent to the eastern shores of  Loch Assynt presented us with a sweeping panorama and we gazed in awe as we drove past the dramatic relic of ancient Ardveck Castle, rising from the still waters, a reminder of the  tempestuous period of Scottish history  over a thousand years ago.

Our planned destination that night was The Durness Hotel, on the most northerly point of North West Scotland, where The Atlantic constantly pounds the rugged coastline.  We hastened along the narrow road, with its pull-in places should two cars meet head on. However we drove for an hour without sighting another vehicle, until eventually arriving at Durness, which consists of a small hotel and a handful of bleak cottages.  But the hotel was closed for the season!

It was now late afternoon. The watery sun cast weird hues through the ribbons of sombre clouds, as it commenced its disappearance over the horizon.  The gloaming was settling in and the countryside prepared for sleep. But no rest for us!  A close perusal of the AA Road Atlas left us in no doubt that the nearest village and hotel was Tongue.  This meant a lengthy tedious drive along a single track roadway that curved torturously along the northern coast before almost circa-navigating the sea inlet – Loch Eriboll – then across the moors to The Kyle of Tongue, and hopefully a hotel that was not closed for the season!

Eventually Tongue was reached, having crossed the spindly viaduct over the Kyle.  We were the only guests at the crumbling hotel, but enjoyed a hearty meal of poached local salmon followed by roast venison in front of a cosy peat fire, plus a hot bath and a sound sleep in an old fashioned creaking bed.

The following morning we made an early start and pointed the nose of the Rover east in the direction of Thurso.  Unlike the vast majority of Highland terrain, the strip along the northern region is relatively flat and almost heath like.  On the approach to Thurso I swung a left turn which took us to the ferry terminal of Scrabster, the place of embarkation for Stromness, the principal port of Orkney.

## ORKNEY.

The Orkney Islands are separated from the Scottish mainland by a stretch of water known as The Pentland Firth.  It has the

*The drive north from Ullapool in the direction of Durness includes some magnificent stretches of roadways, which wind their way through the glens of the Northwest Highlands. Note the poles on either side of the road, designed to indicate the depth of snow during the long winters.*

reputation of being the stormiest water crossing in the world! I had prepared Margery to expect the worst. She tends to suffer from *mal-de-mere* even on dry land!

The *St. Olaf* was obligingly awaiting our arrival (it would seem) alongside the jetty at Scrabster. This cavernous ferry had been custom built to cope with the huge seas that so often prevented conventional ships from venturing out into The Firth. It obviously had an uncommonly deep draft and broad beam, in order to help stabilise the vessel in high seas. Large trucks were accommodated deep down in the bilge area, whilst the lighter cars were directed to the higher decks.

With the car safely stowed away, Margery and I ventured out on the upper deck, where we witnessed a comical sight which had us chuckling. The captain (no less) was leaning over the rail of his bridge with a fishing rod and line, whilst peering hopefully down into the water! It occurred to us that he probably preferred a fresh fish for lunch rather than the cook's frozen variety in the galley!

Margery suspected me of exaggerating when I had warned her of the mountainous seas normally experienced in The Pentland Firth. That day the surface of the water resembled that of a mill pond. The *St. Olaf* quietly surged its way across the flat sea on its journey to Stromness. On our starboard bow the island of Hoy rapidly took shape. But it was The Old Man of Hoy upon which our attention was focused. Rising dramatically out of The Atlantic to a height of 449 feet (137 m.) this needle-like red sandstone stack has been separated by the erosive power of wind and wave from the cliffs of Rora Head. It stands like a sentinel and is considered the UK's most challenging climb.

Stromness is a typical island port, grey and sombre, having endured centuries of huge seas and battering winds. We located a modern B.& B. cottage on a rise overlooking the town where we were warmly welcomed by a cheerful lady with a most captivating Orcadian lilt to her dialect, whose name was Alison Clouston. Importantly she was the repository of all the "goings ons" on the islands.

"Yes of course, to be sure I ken of the tractor collection. After all, everybody knows of Billy Dass out on Bhu Farm" she said pointedly.

I let out a sigh. Thank goodness! We were now on the home run.

Bhu farm is located on The Isle of Burray. This involved us in a scenic drive through rich farming country and clusters of stone cottages. No trees grow on these islands on account of the perpetual winds constantly bombarding the landscape. Indeed the soil is rich in phosphate due to aeons of winds crashing against the cliffs and hurling sea shells the breadth of the land. Barley for the local distilleries is one of the principal crops.

We drove east following the road around the northern shore of Scapa Flow, that historic sheltered naval sanctuary, which features so prominently in the maritime history of both world wars.

A pause for lunch in the main commercial centre of Kirkwall was followed by a quick tour of St. Magnus Cathedral, a majestic edifice in pristine condition dating back to the twelfth century. Then it was south to the Isle of Burray, which included driving across an artificial causeway, part of the Churchill Barriers designed to protect the naval ships in Scapa Flow from U-boat attacks during World War 2.

There are three wee villages on Burray; Northtown, Southtown and Burray Village. A tractor pulling an overfull dung cart was emerging through a field gateway and I inquired from the driver if he knew of Bhu Farm.

"Of course" he retorted, seemingly amazed that anyone would not know of its whereabouts. "Ye canny miss it" and with that he proceeded to wave his arms indicating a farm along the road.

## SUCCESS AT LAST!

The stout oak door was opened and a rotund jolly figure appeared.

"Aye, whit can I do for ye?" he asked inquiringly.

"I am looking for Mr Dass. A fellow along the road said this was Bhu Farm" I explained.

"No person here of that name" he retorted sharply. I was perplexed, the tractor driver was quite positive with his directions.

"Are ye frae the tax department?" he asked suspiciously.

"No, certainly not!" I was taken aback by the suggestion. "I am from Australia. I write books about tractors and am searching for The Bhu Farm collection" I shot back at him.

"Och weel. That's all richt then. You'd better come awa ben. I'm Billy Dass and you and yer wife are very welcome. I'll

*Billy Dass of Bhu Farm on the Isle of Burray, proudly showing off his collection of early David Brown tractors. Note the bleak treeless landscape in the background. No trees grow on the Orkney group of islands, on account of the prevailing gale force winds that continually assault the landmass following their long 4,000 mile journey across the North Atlantic.*

pour oot a wee dram and then I'll introduce ye to my collection of David Broons"

And this is where the story ends. Billy Dass proved himself to be quite a character. We were introduced to his wife and enjoyed afternoon tea, following the inspection of his David Brown tractors.

Did our lengthy search end in an anti-climax? Absolutely not! Was it worth all the time and effort to finally track down Billy's collection? Yes! Have I forgiven Donald for his vagaries? Most definitely! The search encouraged us to travel through some stunningly beautiful country, we otherwise possibly would never have visited.

As for that elusive collection? Certainly so far as tractor collections go, it was a small but significant assembly of David Brown tractors. Billy enthusiastically started each and proudly drove them around the ancient farm courtyard. Their bright Hunting Pink paintwork was a welcome contrast to the sombre grey of the stone steading buildings.

What *is* important, I can now boast of having seen the closest classic tractor collection to the North Pole! How good is that? Not many of my fellow tractor enthusiasts can match such an achievement!

# ONE THOUGHT LEADS TO ANOTHER.

## A RAMBLING HISTORY LESSON.

### THE DRIVE.

Funny how things work out! A year or so ago, as a result of so much rain here on the Mid North Coast of NSW, the gravel drive up to our house from the front gate became fit for only off road vehicles, due to erosion and wash-aways A visiting friend arrived with a gleaming new Jaguar. He took one look at the drive and saw the wisdom of leaving his car at the gate and walking up the hill, rather than have its bottom (the Jag that is) damaged by the deep ruts.

Some major road reconstruction was suddenly elevated to the top of my priority list.

A telephone call to our friendly sand and gravel man resulted in a procession of heaped gravel trucks arriving the following day. Their loads were duly tipped along the length of the drive.

Now it was time to fire up the Landini, our farm work horse which is equipped with a heavy duty front-end-loader. With one eye on a threatening build-up of storm clouds in the southern sky, I roughly spread the gravel heaps, an easy task for the 4 wheel drive Landini.

So far so good. I then hitched the grader blade behind our vintage BMC Mini tractor and proceeded to level and grade the gravel. The 3 point linkage of this brilliant little lightweight tractor is amazingly sensitive, thanks to the Harry Ferguson designed hydraulics. Infact the diminutive orange tractor is the perfect machine for operating a rear grader.

By late afternoon the job was done. I stood back and admired my handy work. The gentle contours sweeping up the hill and around the curve had an almost sensual appeal and I was certain the unequivocal silky smooth surface could have served as an inspiration to our local council engineers.

At precisely 6 p.m. the rain started again. Within five minutes the heavens had erupted and a torrential downpour shook the house, accompanied by a noise resembling a runaway locomotive.

But torrential downpours are supposed to last only a few minutes. This one continued to screech and hammer all through a sleepless night and only started to ease around mid morning!

After breakfast, clad in my wellies and rain gear, I ventured out to survey the damage. First, a squelchy crossing to the rain gauge. My mind at first would not accept the mathematics.

Slowly it registered that, without any question of doubt, over twelve inches of rain had fallen from the sky overnight, and it was still absolutely pouring! But everyone else thinks in millimetres. So get my brain into gear – wow, that is 317 mm.

The house, sheds and cattle yards are all on top of a hill. Accordingly, apart from utter saturation everywhere, there was no damage. But then I went to survey the drive !

As a youth in Scotland I can recall opening the furrows for the planting of potatoes. This was achieved with a deep digging single furrow mouldboard plough mounted behind a David Brown Cropmaster. The depth required was 14 inches.

Upon gazing with shock over the panorama of the remnants of my hitherto manicured drive, I experienced an immediate recall of a Scottish potato field. Because that is exactly what it resembled. Deep furrows gouged everywhere. Of the tonnes of gravel there was little evidence, until I slithered down to the bottom of the hill where I discovered a great mountain of the stuff, completely blocking the gate.

It took me all the next day with the Landini to shift the gravel back onto the drive and reshape it again with the stalwart little BMC Mini tractor. However I certainly was not complaining. Anyone who has been around the Bush for as many years as I, knows never to bellyache about too much rain. The thing is, one never knows when it will rain again. Give me mud over dust any day!

# THE HISTORY LESSON.

In a contemplative mood, I started thinking about what a truly remarkable tractor the BMC Mini really is. Few people, and here I include classic tractor enthusiasts, appreciate the fact that despite being a BMC product (British Motor Corporation and later British Leyland) it was totally designed by Harry Ferguson (Research) Ltd. That is, apart from the diesel engine which was a derivative of the BMC Series 1 petrol engine as fitted to the Morris Minor 1000, etc.

Still in my contemplative mood, it occured to me that the name of Harry Ferguson may not mean much to many of today's younger farmers. Obviously anyone connected with agriculture is familiar with the name *Massey Ferguson*. But

*The BMC Mini tractor, restored by the author. This is the BMC product designed by Harry Ferguson Research Ltd.*

who was Massey and who was Ferguson?

In a nutshell, Daniel Massey created a farm machinery manufacturing business in 1847 at Ontario, Canada. In 1891 his firm merged with its main competitor, A. Harris, Son &

*The 4 cylinder diesel engine of the brilliant BMC Mini tractor was basically the 948 c.c. BMC A Series, as fitted to more than 2,000,000 Morris and Austin vehicles for over a decade. However the engine was converted into a diesel by noted diesel design engineer Harry Ricardo and given a rotary fuel pump and injectors. These special Series A diesel engines were assembled at the BMC associate company Newage (Manchester) Ltd.*

Company, also of Ontario.  Within two decades of spectacular growth, Massey Harris Limited had become the world's largest producer of agricultural machinery.

Harry Ferguson, an Ulsterman born and raised on a farm named Growell  situated near Belfast, became a legend in the tractor world, culminating with the success of his innovative Ferguson tractor and the revolutionary design of the Ferguson three point linkage system.   His method of attaching implements to tractors has been adopted by virtually every modern tractor manufacturer.

In 1953 Massey Harris purchased the Harry Ferguson empire and for just over three years the new conglomerate traded as *Massey Harris Ferguson*. In 1957 the name of Harris was dropped and thus the name *Massey Ferguson* came into being.

## SPEED.

Having dealt with the boring bit, I would now like to introduce another element into this narrative – and that is *speed* and how it led ultimately to the formation of Harry Ferguson (Research) Limited. (This will take us back to the BMC Mini tractor.  Trust me).

Harry Ferguson's obsession with speed first occurred in his youth in 1902 when, at the age of 18, he commenced his engineering apprenticeship with the Belfast firm of Hamilton & Ferguson.  A year later, his elder brother Joe who was a partner in the business, branched out on his own, trading as J B Ferguson & Co., taking Harry with him.

In order to publicise the fledgling company, Harry enthusiastically embraced the new dare-devil sport of motorcycle racing. In these early days motorcycles were dangerous contraptions. (So what's changed?) They were usually equipped with only one ineffective rear wheel brake, belt drive, oil lamps and unsprung girder forks.

Around 1906 Harry's brother acquired the franchise for the outstanding range of Argyll cars, manufactured in Scotland. These technically advanced vehicles were powered by an almost totally silent sleeve valve engine, designed by a Scotsman named Bert McCollum. Interestingly, McCollum also designed the sleeve valve engines fitted to some of the first production Glasgow tractors in 1919.

Harry Ferguson made the switch from motorcycle to car racing and in 1908, driving an Argyll, took first place in the Bangor Hill Climb and a second place in the arduous Irish Reliability Trial.

In 1909 Harry Ferguson's obsession with speed saw him gravitating to the world of aviation. In the attic above his brother's garage he constructed the wings and fuselage of a monoplane aircraft. It is questionable if at this stage he had ever even sighted an aircraft and my research suggests that the sum of his knowledge had been obtained from grainy photographs and newspaper cuttings.

Remarkably therefore, his prototype ash frame and canvas skinned flying machine actually flew on the 31st December 1909.  Certainly it was only for a short duration, but an improved model won for him a £100 prize for the first aircraft in Ireland to successfully fly a three mile course.  He was also the first Britisher to design and fly his own aircraft.

A few years later Harry Ferguson's innovative mind turned to tractors and his unprecedented success in this realm is widely known and respected by farmers around the world.  However in addition to his awakened interest in tractors, he always made time for further involvement with motor vehicle racing.

He successfully lobbied the Ulster Government to introduce the Road Races Act 1922. This opened the way for the initiation of the legendary Ulster Grand Prix and Ulster Tourist Trophy races, both of which became major events on the European road racing calendar.

Years later, in 1956, having sold his tractor interests to Massey Harris, Harry Ferguson (Research) Ltd. was launched. Its purpose was three fold.  (1) The continuance of research into tractor hydraulics.  (2) The fulfilment of a contract with BMC to design a lightweight tractor.  (3) To further develop work that had been commenced by the celebrated racing driver Freddie Dixon, into the research of a 4 wheel drive transmission for race cars.

Harry Ferguson personally supervised the work on the 4 wheel drive designs. The British car industry was amazed when an experimental Ferguson saloon car, fitted with a constant 4 wheel drive system, was demonstrated in early 1957. The styling of the vehicle loosely followed the lines adopted by Triumph in the 1950s, obviously a legacy of Ferguson's earlier association with Standard/Triumph cars.

In 1966 a significant motoring milestone was reached when the Ferguson P99, the world's first 4 wheel drive Formula One race car, was driven to victory on a rain deluged track by Stirling Moss at Oulton Park.

The first production road car to be equipped with the

*A rare 1936 photo of Harry Ferguson driving his first production tractor – the Ferguson Type A, which was manufactured for him by David Brown Ltd., of Huddersfield. (Photo courtesy The Ulster Transport Museum).*

Ferguson constant 4 wheel drive system was the high performance Jensen Interceptor FF. Later Audi and Mazda acquired manufacturing rights and today the Ferguson concept has been widely adopted by car manufactures around the world.

Testimony indeed to the engineering brilliance of a man who is largely remembered only for the little grey Ferguson tractor – and by me for the clever BMC Mini tractor that fixed my drive!

# FROM RUSSIA WITH LOVE??

**THE VODKA.**

A few days ago I washed and polished two tractors, two ride-on lawnmowers and our ostentatious classic red roadster. Whilst never neglecting the roadster (perish the thought) I confess to having for some time disregarded the cosmetic requirements of the tractors and lawnmowers.  As a consequence, a considerable degree of elbow grease was required to remove the several months build up of dirt and putridity.

I realise the foregoing can hardly be construed as earth shattering information.  However it is by way of providing an acceptable explanation why late that afternoon, following my unaccustomed burst of athleticism, I could have been observed reclining on the back patio nursing a vodka and orange.

Now, not being a chap noted for my mental dexterity, the vodka must have stimulated the functional activity of my grey cells for I found myself pondering the question, why is it that Russia produces the best vodka?

This train of thought meandered around for a while until

my tractor blinkered mind took control and posed the ultimate Russian question. Why was the design of the initial Soviet tractors that came to Australia in the 1960s so antiquated, considering by that time Russia had pioneered space exploration with the launch of Sputnik on 4th October 1957?

Suddenly and challengingly, here was a topic for contemplation!

## MY FIRST SIGHTING.

Light years ago, 1969 I think it was, during the period when I had the dubious distinction of being General Manager of Cumberland Tractors Pty. Ltd., the phone on my desk jangled into action. I recall it was a Friday morning and, following a particularly stressful week, I had been pleasantly contemplating a weekend afloat on *Coolalie*, which would be patiently awaiting our pleasure at her mooring on Pittwater.

"Arnold Glass here Ian. Just wondering if you have a few moments to walk across the road? There's something I'd like to show you".

The Cumberland Tractor premises were on the northern side of the Parramatta Road at Auburn (Sydney). Directly opposite, on the original site of The Sydney Meat Works, Arnold Glass had created his vast Capitol Motors complex.

Glad of the diversion, I took my life into my hands and dodged across the mayhem of Parramatta Road and into the glitter of Arnold's showroom.  He awaited me, standing beside a line of half a dozen of the strangest tractors I had ever encountered!  This was my first sighting of Belarus tractors.

I first met Arnold Glass in 1953 at a time when he owned a diminutive yard located somewhere around Sydney's Campbell Street, from which he sold second hand motor cycles.  He convinced me (wisely as it turned out) to invest my meagre savings in a Norton ES2.

But we had both moved on from these early days.  In 1969 he was by then a multi-millionaire highly respected motor industry magnate and I was but a humble overworked and underpaid tractor industry executive.

Arnold explained that he had imported this brace of Russian tractors in order to assess their marketing potential.  He fully appreciated that they would encounter fierce competition from the many entrenched and respected brands that dominated the already crowded sales arena.  Particularly, he acknowledged, from Massey Ferguson, the top selling tractor at that time and for which Cumberland Tractors happened to be Australia's premier dealership.

In short, these Russian machines would *have* to be good!

And good they were not!  Infact even my first cursory glance told me that Arnold's best plan would be to put them back in their box and consign them back to Russia.  I admit it also crossed my mind that Albert G. Simms (the scrappy, who was the founder of that global giant Simmsmetal) was offering a good price per ton for scrap iron.

However I kept these unworthy, but possibly sagacious thoughts to myself and felt that courtliness and decorum dictate I go through the charade of carefully inspecting the tractors.

But Arnold Glass had not achieved his considerable eminence without having developed a razor sharp ability to perceive hesitation and doubt when in the minds of others. He asked me to "Lay it on the line" and speak my mind.

Where to start ? I walked around the tractors and urged him to take another long look at them. Although ranging from around 30 to 60 h.p. (according to my recollection – after all this all happened over forty years ago) they stood tall and gangly mounted on absurdly narrow rowcrop tyres. Their profile was akin to American tractors of the 1930s such as the Massey Harris Pacemaker or the Oliver 70.

The tractors had either air cooled or water cooled diesel engines but with unbelievably rough cast blocks adorned with odd-ball ancillary equipment. Crude looking inline diesel fuel pumps, unfamiliar electrical equipment, worrying air filtration systems, non-standard threads and bizarre electrical connections everywhere, were all suggestive of a lack of modernistic engineering integrity. Or perhaps, the Bolsheviks had simply been concentrating too much on their vodka!

Three of the tractors were equipped with rear linkage of questionable category but the supporting hydraulics were basic and devoid of depth control, draught sensing, or the other sophistications which Australian farmers took for granted.

At Arnold's bidding, I fired up one of the machines and took it for a wander around the premises. The seat was agreeably comfortable but many of the controls difficult or awkward to reach. Depressing the clutch required an extreme effort of physical application that left my leg aching after one application. Selecting a gear was an exercise in concentration, frustration and good guesswork. The steering was seemingly custom designed for the bulging muscles of a Soviet navvy. Even climbing on board was a challenge requiring a level of gymnastics to which I was unaccustomed.

So I laid it on the line, as requested.

It will come as no surprise therefore that Arnold Glass did not pursue his aspiration of becoming a Belarus distributor.

The last time our paths crossed was in 1979 at a dinner party at Gorian Station, Burren Junction (next door to our property Chelmsford) hosted by Jim and Dibby Alker. Arnold was one of the guests and we laughed as we recalled the Belarus experience. Surprisingly he also had recollections from 1953 of the Norton ES2, plus a sleek MG roadster which Margery bought from him in 1972.

# BELARUS.

"Belarus" was simply a trade identification name introduced in 1961 and applied to all Russian tractors and related equipment exported by the Union of Soviet Socialist Republic trade department V/O Traktoroexport. There were literally scores of tractor manufacturing plants in Russia and in their homeland the tractors were sold under their respective brand names.

The Russian tractor industry was spawned in the early 1900s, but became of age in 1924 when American Fordson Model F tractors were produced under licence in Kirov and traded as Poutilovets. Around the world the Fordson was the top selling tractor and proved to be equally popular in Russia. By 1932 50,000 Poutilovets had been manufactured at the Kirov plant.

If the Russian tractors of the 1960s and '70s lacked *savoir-vivre*, their makers certainly impressed the Western World with production figures. By the early 1970s Russian factories were churning out over 1,000,000 tractors each year. The Soviet Block had suddenly become the World's largest producer of tractors !

The 1978 "Power Farming Annual" carried its first mention of Belarus tractors being imported into Australia by V/O Traktoroexport. The MTZ-80 and MTZ-82, the two models announced, were a huge improvement over the units which I had inspected in 1969. They were however still years behind in technical design from any other tractor on the Australian market.

V/O Traktoroexport claimed the two tractors had been tested at the University of Melbourne tractor testing facility. I find this claim perplexing. The highly regarded tractor historian Doctor Graeme R. Quick, in his excellent book "Australian Tractors", lists all the tractors tested under The Australian Tractor Testing Scheme and there is no mention of any Belarus?

Over the intervening years an expanding range of Belarus

*Belarus introduced the T–150K, 165 h.p. four wheel drive heavyweight tractor into Australia in the late 1970s. Apart from the uncomfortable seating arrangement and driving position, this was in fact a well designed tractor produced at the Kharkov Tractor Works. Its V6 turbo charged diesel engine acquired its 165 h.p. at 2,100 r.p.m. In 40 years the Kharkov Works produced 1.6 million tractors, most of which were sold to farmers in Eastern European Communist influenced countries. (IMJ archives)*

tractors have been progressively introduced to Australian farmers, including the giant K701 powered by a massive V12 turbo-charged 300 h.p. diesel, weighing in at an unballasted 13,027 kg. But reliability problems persisted to tarnish the reputation of the Russian tractors. It was not until the 1990s that these were largely overcome.

I am certain the Belarus problems stemmed from a lack of quality control exercised by factory management. I have had personal experience of this incompetency exhibited by other Eastern Block tractor manufactures, plus also cars, trucks and earthmoving equipment produced in Socialist factories. Production targets per month was the doctrine of all factory bosses and was vigorously enforced by their political watchdogs. Thankfully these days are long gone.

Today, modern Belarus tractors are produced in state-of-the-art factories and are undoubtedly world class.

Er – why is it the Russians make the best vodka?

# JOHN DEERE – THE FACTS!

## LOGIC.

I have not checked the statistics recently in order to establish precise figures, but one doesn't require a degree in agricultural economics to be aware that John Deere tractors enjoy a very healthy share of the total tractor market in Australia. It is impossible to motor through any of our broadacre farming districts without sighting at least a few of these green and gold machines toiling away.

Having personally owned and farmed with John Deere tractors, I can understand the reason for their popularity. However I must hasten to add, I do not intend this article to be a promotional advertisement for Deere and Co.! I mean to say – all tractors are pretty good today. Well, aren't they?

I am merely thinking that as there are so many J.D.s around, it therefore logically follows that legions of farmers own them, and according to my somewhat paroxysmal syllogistic reasoning (phew) it is justifiable to assume that a significant percentage of these said owners would at least be mildly interested in becoming acquainted, that is assuming they are not already so,  with precise and accurate information relating to the origins of John Deere tractors, upon which they could contemplate during their ensconcement within the air conditioned confines of their tractor cabins, as they diligently navigate the big machines around the workings of their respective paddocks.

Gee – did I write that?

(Before proceeding, please note – I was not involved in the script writing of Sir Humphrey's lines in 'Yes Minister'!)

## MISINFORMATION BY ALLEGED 'EXPERTS'.

At the risk of being labelled a 'smarty', believe me when I state categorically that the majority of tractor scribes get it wrong, when explaining the genesis of John Deere tractors. The frequent misconception is that Deere and Co. of Moline, designed its first tractor in 1918 and named it Waterloo Boy.

Not so!

So how come I should know better and what makes me right?

Simple really.  Back in the Nineties I was invited to spend as much time as I wished, exploring the Deere and Co. archives, at Moline, USA.

During the visit, my endless requests for information and the inspection of specific documents were greeted with extreme tolerance and good humour. I am grateful to Les J. Stegh, Senior Archivist, and his competent staff for their warmth and cooperation.

So there you have it! The reason for being virtuously confident of my facts, is that I performed indepth research into my subject, instead of cobbling information from other's writings, as regrettably is the custom of many 'modern' alleged tractor historians and writers.

## THE TRUE FACTS.

During the first decade of the 20$^{th}$ century, the Deere & Co. management team observed with some anxiety that a number of their farm machinery competitors were becoming involved with tractors. They were particularly concerned that brands such as International Harvester, Sawyer Massey, Hart Parr, Rumely, Minneapolis and Case would expand, to the detriment of John Deere, which at that time had not entered the tractor arena, and thus would be left behind with the new technology.

Accordingly, in 1910 the company negotiated an arrangement with the Gas Traction Company of Minneapolis, to distribute its tractor known as The Big Four. The mammoth tractor, weighing around 10 tons, was designed by D. M. Hartsough and had been successfully marketed since 1904. It was so named as it was claimed (wrongly) to be the first American tractor powered by a 4 cylinder engine. However, Deere's association with The Big

Four failed to return the profits anticipated and the agreement endured only for a short period.

As a consequence, in 1912 the President of Deere & Co., William Butterworth, gave approval to his senior engineer, C. H. Melvin, to proceed with all due haste, the design and production of a medium-weight tractor. The project took two years and despite the incorporation of many engineering innovations, was finally abandoned, to the dismay of management, following unsuccessful attempts to overcome mechanical problems with the 3-wheeled prototype.

The Vice President, Joseph Dain, took it upon himself to pursue the tractor project and succeeded in convincing the board to allocate the substantial funding required for the commencement of an entirely different tractor design.

In 1916 Dain's all-wheel-drive John Deere tractor went into production. The first of these 3-wheeled machines proved to be underpowered and subsequent units were equipped with a Mac Vickers designed 4-cylinder engine, which produced 24 belt h.p. resulting in a satisfactory 3,000 pounds drawbar pull (at an unstated speed). The tractor was equipped with a complicated 2 speed transmission, which could be changed on the move under full power.

But still there were problems. The price of the Dain John Deere 3-wheeled tractor proved uncompetitive, largely owing to the complexities of the all-wheel-drive design. Only an estimated 200 units were produced before the project was discontinued.

With a degree of urgency the board looked around for an alternative tractor to market. They focused their attention upon the Waterloo Gasoline Engine Company, which had been manufacturing Waterloo Boy tractors at its plant in Waterloo, Iowa, since 1913. H. W. Leavit, the firm's chief engineer, originally designed a tractor named the Big Chief for a rival company. Upon joining the Waterloo Gasoline Engine Company, Leavit simply redesigned the Big Chief, but with numerous improvements, and thus created the Waterloo Boy.

In 1918 Deere & Co. purchased the Waterloo Gasoline Engine Company, together with the Waterloo Boy tractor and the services of H. W. Leavit. By this time over 8,000 Model R and N twin-cylinder Waterloo Boy tractors had been purchased by farmers around the world, who were more than pleased with their new tractor. Some were exported to Britain, where they were marketed as the Overtime Model N. (Interestingly, Ulsterman Harry Ferguson became their distributor in Ireland.)

Despite the Deere and Co. acquisition of the Waterloo Boy design, it was decided to continue to sell the tractor under the brand name of Waterloo Boy until 1923, when the all new John Deere Model D was released. Significantly, around six years had elapsed since Dain's 3-wheeled tractor went out of production.

The twin-cylinder John Deere Model D became a legend. It was commercially one of the most successful tractors of all time. It also had the longest production life of any model of any tractor. Including several upgrades, it was produced from 1923 until 1953. During this period, nearly 200,000 were manufactured.

Throughout the 30-year production life of the Model D, and up to 1960, a comprehensive range of other twin cylinder John Deere tractors was produced. Included were row crop, broadacre and orchard tractors, also crawlers and industrial units, plus diesel variations.

In recent years, twin cylinder John Deere tractors have become greatly treasured by collectors and museums around the world.

# SEVENTH HEAVEN.

## THE 1999 NATIONAL RALLY.

## A CONSUMING PASSION.

On the whole, vintage tractor collectors are a blissful and easy going lot.  Unusual perhaps in these frantic modern times, this state of beatitude is achieved without recourse to funny cigarettes or even a furtive viewing of *Lolita.*  Their "highs" are the results of an ardent love affair with their old tractors.  This passion reaches its consuming climactic pinnacle at vintage machinery rallies, where the tractors are groomed and paraded in the manner of a 1950s bathing beauty quest. Their owners bask in the warm glow of belief that *theirs* will be the cause of great lust and envy in the minds of all other exhibitors.

The meridian of our tractor shows is the Biennial Rally held by the National Historical Machinery Association of Australia. A different venue is selected each two years and the Seventh Biennial Rally was held in March 1999 at the site of the Henty Field Days, hosted by the Henty and District Antique Farm Machinery Club.

Like all good vintage tractor collectors, there was never

any doubt in the minds of Margery and myself that we should attend. The only question was – which tractor should be carted the 1,000 ks. to Henty. In other words, which tractor would cause other collectors to swoon, hate me with envy or simply cry a little. John Deeres are great, Lanz Bulldogs fascinating, McCormicks – well sort of ho hum, but what about a classic German 1951 Orenstein & Koppel V twin diesel?! Allegedly only four remain complete and running in the entire world! This would *have* to be the ultimate tractor at Henty.

The Rain!

Weeks ahead of the deadline, the Orenstein & Koppel was dragged from the inner recesses of its shed into the bang, bog and paint bay. I had done all the mechanical things to it a year or so ago so therefore all that was required was to attend to the dings, rust and nearly half a century of body neglect. As always, this took longer than expected but three weeks prior to the rally it was ready for paint.

The double mudguards and floor are one unit and it was beyond my physical strength to remove them – even with Margery helping! So they had to be painted in situe. This presented a tedious complication because it involved fiddly masking. You see the engine, axles and transmissions were to be painted dark brown and the bonnet, guards etc. a light brown. Then for additional fun the wheels had to be sweet cream *and* dark brown.

The trouble was it rained buckets – and refused to stop. Then, like a shining beacon of hope, the sun came out. The light brown paint was sprayed with alacrity and then an hour later a thunder storm pelted rain horizontally into the shed and over the tractor ruining the fresh glossy surface. (Once again it occurred to me that collecting stamps rather than vintage tractors would prove a less demanding hobby.) The tractor would have to be allowed to dry then sanded back and totally redone. Over the next week or so the rain continued to pour. Accordingly on the day prior to our departure for Henty, a freshly painted but decidedly sticky Orenstein & Koppel thumped its way onto the Hino truck.

It rained all the way to Henty. South of Wagga Wagga, having been nearly blasted off the road several times by overtaking low flying B-doubles, we came up behind fellow collector Ian Browning's truck with his rare McDonald on board. We followed his spray hoping he knew where he was going. It turned out he didn't – so we *all* took the wrong turning!

## THE SEVENTH BIENNIAL RALLY.

The rain eased as we eventually arrived at the Rally Secretary's office. The entire site was alive with activity. Over 2,000 exhibits, including around 200 tractors, had to be unloaded, allocated stands and then preened. I surreptitiously gazed around at my brethren tractorphiles to see if any had fainted at the sight of the Orenstein & Koppel.  It remained unnoticed and at that juncture was just another tractor.  Puzzled, I could only reason that everybody was engrossed with getting organised.

During the three day event the Orenstein & Koppel did attract considerable interest and Mr. Kodak would have been pleased at the literally hundreds of photos it inspired. However I was the person who nearly fainted when I saw the other Orenstein & Koppel!! Four in the entire world and two of these at Henty.  John Hassan of Condobolin, with whom I had previously only spoken over the telephone, arrived at the site with his.  I guess it was some sort of world gathering of the Orenstein & Koppel clan.

*The ultra rare 1951 Orenstein and Koppel Diesel-Compressor-Schlepper Model S 32 K.  Originally a rusty hulk when discovered, now restored by the author.*

But neither tractor stole the show.  The Emerson Brantingham Big 4, exhibited by Norm Johnston of Sale (Vic.), was the outstanding tractor exhibit at Henty.  As Chief Judge of the tractor section, I created a special Award of Excellence and Historic Significance specifically for Norm's gargantuan machine.  He restored this rare artefact, following its recovery from Outback NSW several years previously and an expenditure roughly equating the National Debt, plus thousands of dedicated hours of hard graft. It has been reborn so that future generations may gaze upon its magnificence with wonder.

Incidentally, I had a sneaky feeling I had been appointed Chief Tractor Judge simply because no one else wanted the job. Frankly, wading around in ones Wellies and Dryazawhatsit in pouring rain, impartially judging 200 assorted tractors scattered around the extensive landscape, I'll admit was challenging. The thing is – they were *all* good, in fact brilliant in the eyes of their owners.

Saturday was judging day. In the morning everyone loved me.  By the end of the day only the few category winners harboured charitable thoughts towards me.

For the record, my interpretation of best "restored" tractor accords with the Oxford Dictionary – "returned to original condition".  A tractor that gleams with its two pack paint, body filled castings, plated nuts and bolts, rubber carpets and polished brass pipes has not been restored!  It has been taken beyond "original" condition.  It has been *tarted up*.  Well that is what I dared to say during my after dinner speech.  Frowns flashed across a few countenances, but the majority of the attentive audience nodded and murmured "Good on yer".

Yes, it did rain all the way home.  But I was happy in the knowledge that thousands of enthusiasts had after all inspected our Orenstein & Koppel.  Although to be honest, I didn't see anyone actually faint!

# THE GLASGOW.

## SCOTLAND THE BRAVE!

Those who imagine that the Scots spend their entire time living with sheep or casting a fly in some remote trout stream, or tossing cabers (that is when they are not tossing back wee drams) may register surprise when I reveal that these hairy folk do infact have additional talents.

Take for example: Sir Alexander Fleming who discovered penicillin.  John Chalmers who invented the postage stamp. John Baird who invented television. Charles Macintosh who invented the raincoat. William Patterson who created The Bank of England. Kirkpatrick Macmillan who invented the bicycle. John Boyd Dunlop who invented the pneumatic tyre. John Macadam who invented the bitumen road. Also, Alexander Graham Bell who invented the telephone.

And let us not overlook the fact that the largest ship building yards the world has ever seen were at Clydebank. Or that Britain's second highest export earner is Scotch whisky. Or that The Forth Bridge is the greatest engineering marvel of the nineteenth century.

So, following that unashamedly cerebral grand tour of

Scottish ingenuity, it is time to examine yet another example of world leading innovative Scottish engineering.

## THE GRAND PLAN.

William Guthrie, a Scottish Engineer and gentleman farmer, watched with dismay as the giant 8 ton  Marshall started to spin one of its enormous rear wheels.  It was becoming patently obvious that heavyweight tractors were simply not suited to the rich moist red clay soils of The Lothians.  Within seconds the big machine had sunk to its axles and was completely bogged.  Only a team of a dozen Clydesdales would have the grip and power to free the now inert tractor.

Even the much lighter Fordson tractors that started arriving in volume numbers from America, could not successfully pull a plough in the damp northern autumn months without continually bogging.  Accordingly, the majority of the thrifty minded Scottish farmers chose to keep their sporrans tightly closed and persevere with their time-proven horse teams. They considered there was no future for these smelly noisy tractors that so rudely destroyed the tranquillity of the countryside.

But not William Guthrie, who spent long winter evenings sketching ideas for a revolutionary machine that would be capable of overcoming the tedious problems of bogging.

In 1918 a historic meeting took place involving the chief executives of three Scottish engineering firms – The Carmuir

Iron Factory, D.L. Motors Manufacturing Co. and John Wallace and Sons.  These well regarded companies had been approached by Guthrie and were impressed by his prototype blueprints.

An agreement was drawn up whereby the three organizations would pool their resources and produce an indigenous Guthrie designed Scottish tractor, which they were convinced would be capable of handling the formidable sticky soils of the region's arable farmlands.

The developmental work was performed in a workshop at Motherwell, Lanarkshire.  Impressed by the first prototype example, the directors of the new company, Wallace (Glasgow) Ltd., exhibited an imprudent enthusiasm for the project. They rushed to acquire a disused munitions factory at Cardonald, an outer Glasgow suburb, in order to commence full scale production of the new Glasgow tractor and envisaged that within a few years 5,000 tractors would be produced annually.

It would appear that no due diligence was applied to a proper marketing assessment program, giving consideration to costings and the competition from low priced albeit inferior tractors, such as the Fordson and Overtime.

## THE GLASGOW.

Guthrie's creation was indeed radical.   He designed the tractor as an *all-wheel-drive* three-wheeled machine.  It featured two front wheels and a third centrally located rear wheel. In effect, no wheel followed in the track of another. Each time the tractor moved forward it was propelled by three wheels all operating on fresh ground. There was no differential in the front axle, therefore a single wheel could not slip or spin. All three wheels would have to spin together – an unlikely event.

In place of a differential the front wheel hubs were fitted with a pawl and ratchet system. This permitted smooth turning by allowing the outside wheel to speed up whilst the other continued at the sustained speed in relation to the rear wheel. Interestingly though, in reverse gear the pawls disengaged the front wheel drive so the Glasgow became a *one-wheel-drive*!

*The author driving a 1919 three-wheeled all-wheel-drive Scottish manufactured Glasgow, which has been restored by Newton Williams and is on display at the Pioneer Settlement Village, Swan Hill, Victoria. (Photo M. Daw)*

Initially sales of the new tractor proved disappointing. Possibly it was too unconventional for the conservative minded canny farmers.

In a bold publicity initiative, a Glasgow was loaded onto an Albion lorry and carted south, across the border to Lincolnshire, where the 1919 Lincoln Tractor Trials were to be staged. This was a major event organized by the prestigious Royal Agricultural Society of England, attracting farmers from all over The British Isles and beyond.

The 800 acre site consisted of moist clay soil, which equated the challenging soil types encountered in The Lothians, Fife and Perthshire. The tractors were categorized according to their horse power ratings. The Glasgow being 26.7 h.p., was placed in Class 2 alongside well established brands including Hart-Parr, Case, Twin City, Saunderson, Peterbro, Emerson Brantingham, Fiat and New Simplex.

The tests were exhaustive and considered every aspect of a tractor's performance. These included drawbar pull, economy, reliability, value, user friendliness (although the term had not been coined back in 1919) and ability to handle clay soil!

Guthrie amazed the spectators when he drove the Glasgow up a 1 in 1.7 grade. The tractor also recorded the highest pull to weight ratio of any tractor in any class. In pulling terms it was considered the equivalent of a 10 horse team.

The judges were unanimous. The supreme gold medal was

awarded to the Glasgow! It rated well beyond that of any other entrant.

Impressed with the result, a London based distributing organization, The British Motor Trading Corp. Ltd., convinced the Scottish manufacturers to appoint it as sole distributor and undertook to purchase the first 25,000 tractors, payment in advance.

Scottish businessmen are noted for their astuteness. They are also normally a wee bitty suspicious of any generous offers extended to them by southern Sassenachs! It can only be assumed therefore that the directors of Wallace (Glasgow) Ltd. had been enjoying a few too many tipples of Johnny Walker Black Label, when they accepted at face value the offer made by the London firm.

Disaster followed! The factory commenced working around the clock in anticipation of the promised volume sales. With only 200 units produced, a telegram arrived from London advising that The British Motor Trading Corp. Ltd. had gone bankrupt – without any money having been paid to Wallace (Glasgow) Ltd. The Scottish firm was devastated and never recovered from the blow. The Cardonald plant closed its doors in early 1924.

Upon reflection, there obviously were other factors instrumental in the financial collapse of Wallace (Glasgow) Ltd. The tractor was brilliant, so the blame must therefore have been the result of ill-considered management financial controls, based on an unrealistic marketing potential.

Over capitalisation of the Cardonald plant certainly would have been a major contributor to the collapse. The 25 acre site included twelve acres under roofing and was budgeted to produce £5.000,000 of revenue within the first five years of trading.   A totally quixotic figure.

The engineering integrity of the tractor is what one would have expected from a Scottish engineer.  But the totally uncharacteristic lack of financial discipline exhibited by the Board was astonishing and will always remain a blot on the reputation of Scottish thrift and financial prudence.

# THE LATE NON-LAMENTABLE 23C.

The vast tractor conglomerate Massey Harris Ferguson Ltd., was created in 1954 as a result of a merger between Harry Ferguson Ltd., UK and Massey Harris Ltd., Canada.

Three years later, in December 1957, the Company decided to shorten its name into the more manageable Massey Ferguson Ltd. (Regrettably this relegated the name of Harris into the dusty realms of the history books and to be forgotten by most. Therefore as tractor enthusiasts, we should remember that Alanson Harris shared his wealth and ingenuity with Hart Massey, way back in 1891 to form Massey Harris Ltd., one of the all time great farm machinery manufacturers.)

From 1958 on, it was decreed that Massey Ferguson tractors would appear in their new attractive livery of red and grey.

Under the presidency of Canadian born Albert Thornbrough, Massey Ferguson sales were booming around the world. Innovative new implements, machines and tractors from both sides of Atlantic were regularly introduced to farmers, who enthusiastically embraced the products.

Plans were in place for a range of Massey Ferguson earthmoving and construction machinery  to be produced in a newly acquired modern factory based at Aprilia,  located 50 kilometres south of Rome, Italy.  Quite separately but equally significantly, the long established Italian tractor firm of Landini would shortly be brought into the M.F. camp.  Acquisitions were also occurring in France and elsewhere.

The future therefore looked encouragingly bright for Massey Ferguson in the late 1950s.

There was however a serious irritant which presented a constant concern for the Massey Ferguson board.  The diesel engine fitted to the top selling Massey Ferguson tractor – the M.F. 35, the successor to the Ferguson 35 – was attracting an unacceptable degree of adverse criticism from wherever it was sold.

The 4 cylinder diesel engine was made for Massey Ferguson by Standard Cars Ltd., of Coventry, England.  Identified as the Standard 23C, it was designed by the high profile English diesel engine authority Sir Harry Ricardo, founder of Ricardo & Co. (1927) Ltd., pioneers of the Vortex and Comet diesel combustion principles.

The 23C had pre-combustion chambers and indirect injection.  With a capacity of 137.8 cu. inches it developed

37 b.h.p.  The compression ratio was a relatively high 20 to 1.  But there were gremlins in the system!  Somewhere in this configuration Ricardo had not got it quite right.  *The engine could prove frustratingly difficult to start on a cold morning.*

Yes, I am aware that there are many owners and past owners out there who will dispute my comments regarding the starting character of the 23C.  The indisputable fact remains.  The concern Massey Ferguson had with the problem was so great that in 1959 it was the main trigger for the Company's purchase of F. Perkins Ltd., of Queens Street, Peterborough, England, for the bargain price of £4,500,000.

In one clean sweep, Massey Ferguson now owned the largest manufacturing facility of diesel engines in Britain, and importantly was immediately able to replace the troublesome Standard 23C in the M.F. 35 with the vastly superior Perkins 3-152  3 cylinder power plant.  The updated version was released in late 1959 and remained in production until 1962.

Although not as smooth running and sweet sounding as the 23C, the 3-152 incorporated direct injection and was a willing starter in all climates.  The increased torque advantages were also immediately apparent.

The new 3 cylinder diesel M.F. 35 evolved in 1962 into the 35X and later into the 135, which became the world's top selling tractor in the 1960s.

Returning briefly to the question of the 23C engine and

its starting idiosyncrasies, it is true that in Australia with our warm climate, the problem was not as acute as in other less temperate climates around the world. Never-the-less the introduction of the Perkins 3-152 brought a resounding collective sigh of relief from M.F. dealers throughout our country.

To conclude the story of the Standard 23C, this epistle would not be complete without revealing the fact that Massey Ferguson achieved a remarkable marketing accomplishment when it sold its entire remaining unused stock of the 4 cylinder

*This beautifully restored Ferguson 35, in its grey and bronze (commonly but incorrectly generally referred to as 'gold') livery, was photographed at The 150th Massey Expo held in New Zealand. Clearly evident is the Standard 23 C 4 cylinder diesel engine, which in cold climates could prove very difficult to start in the early morning.*

23C engines to its competitor – Allis Chalmers.

The Allis Chalmers Corporation of Milwaukee, USA., owned a subsidiary tractor plant located in the English town of Essendine, Rutlandshire. Its main purpose in life was to produce Allis Chalmers Model B tractors for the British, European and Commonwealth markets. This enabled the tractors to be sold through "soft" currency channels, as distinct from "hard" currency, i.e. expensive U.S. dollars. They were also automatically eligible for Empire Preference tariff rates in Australia and New Zealand.

However the sales of Allis Chalmers tractors had gone into decline. The opportunity to acquire an immediate bulk supply of "Massey Ferguson" diesel engines seemed like manna from heaven to the management team of the struggling British Allis Chalmers facility. This was a clearly somewhat naive management team that had obviously not done its homework and as a consequence, failed to be aware of the engine's short-comings!

The Allis Chalmers' Product Design Department was pressed into producing rushed drawings for a replacement tractor that would utilise the newly acquired engine. This, it was hoped would revitalise Allis Chalmers tractor sales. As early as November 1960 the new Allis Chalmers ED 40 was announced to the world – powered by the ignominious Standard 23C engine.

The ED 40 (English Diesel 40 h.p.) was perhaps the only dud tractor produced by the Allis Chalmers Corporation during its long history.  Volume sales failed to eventuate and fell well short of expectations. British farmers had become shrewd buyers and preferred to invest their hard earned savings into such familiar and well tested icons as Fordson, David Brown, Massey Ferguson, International, etc.

In a desperate effort to improve sales, 450 of the ED 40 units were exported to the USA  Despite the introduction of a clever Depthomatic draft response control, North American farmers, like their British counterparts, were not enamoured by the Allis Chalmers ED 40.  In addition to the inherent engine problems, structural failures occurred with the 3 point linkage arms and drawbar.

In 1966 production of the ED 40 was discontinued. The board of Allis Chalmers considered that it was no longer financially viable to continue the operation of the Essandine plant.  Accordingly, no doubt to the delight of Massey Ferguson, Allis Chalmers withdrew its presence as a manufacturer from the British tractor scene.

# WARM MEMORIES ARE NOT COOL!

It certainly is not COOL for an aging tractorman to talk about "the good old days". To youngsters (anyone under 40 years of age that is) these distant halcyon times are as remote to their thinking as the Craters of the Moon.

I note that even my grandchildren exchange meaningful glances when I have the temerity to raise the subject of my glorious boyhood days in Scotland. When I explain that when I arrived in the Antipodes in 1952, Australia enjoyed the highest living standard in the world coupled to the lowest cost of living in the Western World, they exhibit only polite interest.

Sir Robert Menzies is frequently lampooned by some of the more inane contemporary talkback radio jocks, as an autocratic royalist hangover from the Colonial era. Yet, we of the bush who were privileged to be there at the time, knew him as an indulgent father figure under whose leadership we all prospered. These were the golden years in Australia.

Despite widespread impressions to the contrary, farming was a more spiritually and meaningfully rewarding lifestyle in the 1950s than in modern times.  Global warming, over zealous greens, indecipherable punitive government legislation and one sided politically motivated and sometimes questionable so-called free trade agreements, were thankfully unknown in these days.

Certainly, in the 1950s farming was more physically demanding.  Bulk grain handling had not yet arrived and farmers were condemned to the back breaking task of humping three bushel wheat bags.  Air-conditioned tractor cabins had not been contemplated. The majority of roads in the country remained rough gravel obstacles, guaranteed to loosen the nuts and bolts of any vehicle.  If the bank loan would not extend to one of the recently introduced Holdens, then it was a choice of either something like an Austin A40 Ute or possibly a 1930s Chev.

However on the whole, farmers and pastoralists were a moderately prosperous lot.  Thanks to the Korean conflict, wool was worth a pound per pound.  Sales of grain and beef (mainly to Great Britain) earned good export currency.  Farm inputs were low in relative terms and returns generous.  Of course farmers of the day would never have conceded this point.

Rural tractor dealers in the 1950s were on a roll!  There was a constant clamouring for new tractors.  Indeed some of the more popular traditional makes were in short supply.

For example, a grain grower who had been waiting months

for delivery of an International AW6 would in frustration likely accept an offer of the more readily available Farmall AM, basically the same tractor as the AW6 but in rowcrop configuration. Interestingly, the tall wheeled rowcrop tractor when fitted with a full width front axle, in field operations proved a better plough tractor than the designated broadacre AW6.

Regrettably, the demand for tractors encouraged some importers with little or no tractor background, to introduce into the Australian market a range of badly designed lightweight tractors that mainly originated from England. These inept often dangerous machines included such brands as Brockhouse President, Newman and Ota, the latter a product of The Oak Tree Appliance Company.

Fordson Major tractors enjoyed volume sales and were in fact in plentiful supply thanks to Britain's vigorous export policy. Whilst certainly not as sophisticated as the North American tractors, the original E27N Major was a capable hard working unit. The new look Major, which arrived in 1953/4 was however an outstanding tractor of the period and, like the Ferguson, many are still in constant use more than half a century later.

Fordson Major sales were considerably increased by being the preferred tractor of the various banks responsible for administering the Commonwealth Government's funding for

Soldier Settler and Public Ballot farmers. The Major was the lowest priced tractor in its horsepower range and if a more expensive make was desired by the new farmer, then the government loan advance would not cover the additional cost. So in the vast majority of instances a brand new Fordson was the farmer's ultimate, but sometimes reluctant, choice.

Tractor implement sales also ran hot as farmers could no longer make do with converted horse drawn antique machinery. Three point linkage had arrived, requiring specialist implement attachments. On the grain farms out on the plains, higher horsepower tractors could pull more and travel faster.

*This Case S-EX, restored by the author, was originally designed by the J. I. Case Company as an export tractor for the Scandinavian market, but prevailing political circumstances determined that other export markets had to be found for this excellent medium sized tractor. Accordingly a number were sent to Australia.*

Accordingly, wider more robust machinery was the order of the day. Implement manufacturers such as David Shearer of Burwood, NSW, Rawling & Co., of Coburg, Vic. and Mitchell & Co., Pty. Ltd., of West Footscray, Vic. were happily obliged to operate two shifts a day, in an endeavour to meet the demands for their range of tillage implements.

Machinery dealers could barely keep up the supply of electric lighting plants, milking machines, Victor lawnmowers, fencing materials and irrigation equipment. Firms such as Dangar Gedye & Malloch Ltd., Grazcos Co-op Pty. Ltd., Buzacott Ltd., and even the traditional wool firms that had diversed into farm machinery, including Elder Smith, Goldsbrough Mort and Farmers & Graziers, were all flat out selling gadgets for the farm.

Tractor company reps. were in the job primarily because of their passion for tractors and a sound knowledge of farming. Certainly, they also had to know how to tie a tie, use the correct knife and fork, be able to organise field days and demonstrate a tractor with a high degree of efficiency – and write out an order.

Being a tractor rep. was considered not only a plumb job, but also a lot of fun. We were free agents and managed to perform our responsibilities capably, without being shackled to head office by that modern day umbilical cord – the mobile phone.

Then there were the perks, including a company car, an expense account and the opportunity to travel around the country at a time before speed limits were introduced on country roads. This did not mean we all drove like maniacs. But we had the luxury of being able to concentrate upon what was through the windscreen rather than what was on the speedo! We knew intuitively what represented a safe and comfortable speed for our vehicle under the given prevailing circumstance. That could be anything from 10 to 80 miles per hour.

Should there be any young readers who have managed to stay the distance thus far, I ask them to note the following.

Way back in the prehistoric 1950s, at the Lanz headquarters in Sydney, providing a spare parts order was telephoned or telegrammed through by 4.30 p.m. it would be sourced from the shelf, packaged and delivered to Central Railway by 6 p.m.

There, it would be sorted and put in the guard's van of the appropriate *steam* train (e.g. The North West Mail, The Riverina Express, etc.) which thundered through the night to such remote places as Burren Junction, Weethalle, Moombooldool or perhaps Gurley. The train would pause at even the most isolated unmanned siding, enabling the guard to deposit the package on the platform.

Therefore the farmer who had suffered a breakdown with his tractor the previous afternoon, would have his spare part by breakfast time the next morning.

I doubt that in this stressful computer driven era in which we all live, there would be one tractor firm or courier service that could match this efficiency. *And that was in the days of steam trains!*

My current observations, which I admit are unashamedly biased, are that tractor company reps. (sorry, area supervisors) with their obligatory lap tops, sales performance charts and polished shoes, pale at the thought that one of their dealers might require them to visit an *actual farm*! I mean, farms are either terrible dusty or horrible muddy places and, well, one has to think of one's shoes!

*Yet another version of the Fordson E 27 Major was the half track, fitted with track gear produced by Roadless of Hounslow, Middlesex. Originally designed for the Welsh uplands, the half track version found limited sales in Australia. This unit – restored by the author.*

# PLOUGH AND BE COUNTED 2.
## A CLIFF HANGER!

## THE CHALLENGE.

All roads led to Cootamundra, during the 2004 Easter weekend,
and along these highways and bye-ways travelled never ending
convoys of every conceivable type of trailer and truck, all
loaded to the hilt with a colourful array of old tractors.   An
old-timer gaping at the continuing procession as it wound its
way through his town or village, could have been forgiven for
thinking he was witnessing the rebirth of  Worth's Circus.

If fact, the occasion to which the old tractors were heading
was *Plough and Be Counted 2,* the attempt to regain for
Australia the Guinness World Record for the greatest number
of tractors ploughing in the one paddock simultaneously.

We Aussie tractormen (and women) felt fairly smug, when
in 2001 we captured the record from the South Africans,
by mustering 298 tractors in a paddock at Yass.  Such a
performance, we felt, could never be bettered.  How wrong we
were!

The following year South Africa well and truly wiped the smile off our collective faces when it retook the record with an amazing 730 tractors.  Wow - *that* would take some beating!

A mere four months later, the Land of the Shamrock and Foaming Black Guinness stunned us all by amassing 1,832 tractors, thus easily snatching the record from the South Africans. An astonishing number of tractors indeed, but it was also incredible  that Ireland could find a paddock sufficiently large!  And believe it or not nearly half of these tractors were grey Fergies.

So the stiff upper lip Brits had an obvious obligation to put these wild Irishmen in their place.  They would show the world how it should be done, by jove.

In August 2003 tractors started pouring in to the Hullavington Air Base in Wiltshire.  Despite Union Jacks flying from the mast heads of noblemen's castles throughout the Realm, summoning tractor drivers from every corner of the Sceptre Isle, only 771 tractors arrived on the scene.  Mind you – 771 is not a bad effort but a long way short of the record set by the Irish.  Perhaps the difference was, the Irish tractormen were each promised a free glass of Guinness?

It seemed therefore that it was up to us in the Land of Oz to set a definitive record that would stand the test of time.  Vic Muscat and Brian Sainsbury took the initiative to repeat their fine efforts at Yass by organising our second challenge, this time against Ireland, and naming it *Plough and Be Counted 2.* George W. Bush would have been proud of us – we had declared war upon Ireland (in a friendly amicable way of course )!

## EASTER FRIDAY.

I arrived at the Cootamundra site late Friday to a scene of utter chaos. Tractors of all descriptions, as far as the eye could see were being off-loaded, parked, driven around, fixed, polished, cursed, fuelled or simply abandoned. The truck parking area resembled a giant Californian used truck lot. In the camping area folks were setting up, cooking, complaining, congratulating, but most of all – yarning!

The official registration tent was staffed by a squad of overworked, overstressed and greatly harassed ladies, each of whom should be awarded Australian of the Year in recognition of their dedication, patience and good cheer. Somehow they dealt with an endless queue of tractor drivers, anxious to register and be told where to go (politely no doubt). A pretty lady in red with a charming smile but the stoicism of a New York Cop, ensured that each tractor was registered prior to entering the sacred four kilometre circumference arena.

In the centre of all this mayhem were the two masochists responsible for the whole affair. Vic Muscat, weighted down with a forest of walkie talkies and mobile phones – all seemingly in use and demanding his attention, whilst at the same time throngs of agitated tractor drivers barraged him with shouted questions. His compatriot Brian Sainsbury scooted from place to place upon his overworked quad-drive in an endeavour to create some semblance of order.

My assignment was simple. I had been entrusted with providing a running commentary of the tractor pull competition on the Saturday and the big event on the Sunday. A scary looking scissor action cherry picker was provided specifically for me, guaranteeing to inject a nervous tremor into my Fife accent. *I don't like heights!* My microphone was connected to a battery of amplifiers aimed at the anticipated legions of paying spectators.

## SATURDAY.

By Saturday morning around 1,000 tractors had been admitted into the arena by the Red Lady. Trucks continued pouring into the unloading area at the rate of one a minute. Spectators also arrived in droves, even although the challenge would not take place until the following day.

The vintage tractor pull competition commenced around 10.30 a.m. There were two sleds and the crowd watched in fascination as thumping Field Marshalls and Bulldogs competed against roaring G.M. powered Chamberlains and purring Internationals. I did my best to provide an informative account of the proceedings. It is perhaps as well that the operators of the three screaming super modified feral tractors were unable to hear my somewhat caustic comments aimed at these ear shattering monsters, which were utterly destroying the ambience of the morning. I confess to joining the crowd in applauding when one particularly obnoxious machine failed to move the sled from its base.

## EASTER SUNDAY.

Sunday dawned full of expectation. Yet another glorious sunny day. Glorious unless you were one of the local farmers praying for rain to break the drought. The arena now contained around 1,500 tractors. But to take the record from the Irish we required 1,833. By noon we were still short of about 200 contenders. It was unthinkable that having come this far we Australians would fail in our challenge.

At 1.30 p.m., from my perch aloft in the cherry picker, I appealed to the dense throng of spectators for any of the many farmers present, who had tractors in their sheds, to dash home and bring them to the site. There were hundreds of unloaded trucks available for the transportation. I announced that Vic and Brian had decided to delay the start to facilitate the hoped arrival of these desperately needed tractors.

A Toyota pick-up was used to shuttle willing volunteer farmers to the truck parking area. I am certain a world record was achieved for the greatest number of farmers ever crammed into the rear of a Toyota!

The situation was tense. Was it going to be a case of so near but yet so far? However, by 2 p.m. a steady stream of returning laden trucks could be seen speeding to the unloading area.

Around 3 p.m. a tractor drivers' briefing was called. A count was taken and I was able to announce to the anxious crowd that finally we had the numbers, but only just, and what if there were some breakdowns or non-starters?

At 3.15 p.m. the drivers emerged from their briefing and like an army of ants strode purposefully as they fanned out across the paddock heading for their awaiting tractors.

I had previously advised the anticipating spectators to start shooting off film the moment the tractors started to move, as dust would soon obliterate the far extremities of the field. But no one was more surprised than me by the amount of dust

*They're off! 1,897 tractors all ploughing in the one paddock at the one time. A Guinness World Record!*

created by the tractor drivers en-route to their machines, and not an engine had yet been started!

Confusing figures from the control centre were being relayed to me by Ron Keech with whom I shared the cherry picker and who was in radio contact with the centre. However these figures all exceeded the required 1,833 tractors to defeat the Irish. I could not have functioned efficiently without Ron as he was the direct contact with Vic and Brian. Ron is no lightweight and occasionally he would move to the other side of the cherry picker platform thus causing our precarious elevated view-point to sway alarmingly. I hoped that somebody had checked the hydraulic lines plus all the nuts and bolts. As mentioned, I have a poor head for heights.

## SUCCESS.

Shortly before 4 p.m. the flashing lights of the attending fire trucks signalled the commencement of the challenge. The ground shook, the thunder from the multitude of exhaust pipes shattered the quiet rural tranquillity, and a dense column of red dust dimmed the sun. The sound could be clearly heard back at Cootamundra, five kilometres away. The packed crowd of spectators clapped and cheered and some even danced. Cameras clicked and flashed. Grins extended ear to ear on the countenances of the tractor drivers, many of whom also waved and cheered.

Everyone present was privileged to experience a once in a lifetime staggering occurrence. The word came through that *a record breaking 1,897 tractors* successfully traversed the paddock simultaneously pulling an assortment of soil engaging implements.

*We Aussies had done it!* The nation now owed a debt of gratitude to the organisers Vic Muscat and Brian Sainsbury, the Mayor of Cootamundra Paul Braybrooks and his team, Mr. R. White who kindly made the land available and of course the tractor drivers and importantly all those dedicated folks behind the scenes who made it all possible.

*The view from the cherry picker. A close inspection shows the late afternoon shadow with the author and Ron Keech aloft. The tractor drivers were elated and just kept going round and round, it seemed as if for ever!*

# TAILPIECE.

I was certainly pleased to have played a small part in such a massive undertaking, and it was just great meeting-up with so many long-standing tractor collector acquaintances. The support from the local and also national media was even greater than expected.

All profits over expenditure were allocated to a number of worthy local charities. This adds significantly to the pleasing results of the challenge.

So a great historic occasion, a great buzz and a great Australian achievement.

# THE FARMALL STORY.

## THE EARLY DAYS.

The original agricultural tractors, which emerged in the early 1900s, were considered as being merely self propelled vehicles designed to largely replace the draught horse or oxen. They were vast in size, noisy, smelly and frequently dangerous. When there was no cultivating to be done, they were simply put away in the shed.

But the old steam traction engines, which the tractors with their internal combustion engines replaced, could also drive a variety of farm machines (such as a thresher) using an endless belt.  So the design guys were put to work and soon belt pulleys became a common attachment on tractors.

However, for most farmers of the era a tractor remained an expensive luxury with limited functions.  Horses were still required and back-breaking manual labour persisted as a way of life.

Apart from the Henry Ford organisation, International Harvester was the world's largest producer of tractors during the second and third decades of the 20<sup>th</sup> Century. Within its drawing office, International had a secret weapon! His name was Bert Benjamin.

In 1924 International Harvester introduced to the world's farmers a *multi-purpose* tractor, which manifested the results of years of research by Benjamin. For the first time, here was essentially a rowcrop style tractor that could not only till the soil and power ancillary machinery with a belt, but could also perform the tedious menial tasks normally carried out by teams of manual labour (often children and women) – such as hoeing, weeding and planting. The new tractor was appropriately named the International *Farmall.*

The Farmall was an instant success. Farmers welcomed the tractor's versatility and saw value in investing in a tractor capable of being used throughout the year and which took much of the drudgery out of their lives.

## THE FARMALL EVOLUTION.

The original production Farmall model in 1924 was designated the *Regular.* Priced at around $500 it was not cheap, particularly when compared to the volume selling Fordson F, which sold for less than half the price. However the perceived savings in labour costs and the reliability of the Farmall, which was becoming evident, convinced 75,386 farmers to purchase a Farmall during its first five years of production. Indeed many

Fordson owners were persuaded into trading-in their tractor against the more functional International Farmall.

The little 4 cylinder over-head valve removable sleeve engine was a magnificent example of a simple power plant that was the essence of reliability! With a displacement of a mere 220 cubic inches, it developed 18.03 belt h.p. at 1,200 r.p.m., which was adequate for the many tasks of which it was capable. Three forward speeds were provided which represented a top speed of 4 m.p.h. (It should be noted that in the late 1920s, all tractors were equipped with steel wheels, restricting their top speeds to under 5 m.p.h.).

Over the years, continuing improvements were introduced into the Farmall, plus a wide range of matching implements.

In 1932 the first of the newly released Farmall F12s arrived in Australia. These were a lighter version of the original Regular model featuring a reduced size engine of 113 cubic inch displacement, and developing 16.2 belt h.p. at 1,400 r.p.m.

Modern farmers are frequently puzzled when they learn of the seemingly tiny horse powers produced by early tractor engines. They wonder how a 16.2 h.p. tractor, such as the Farmall F12, could ever have done a proper day's work. For after all, many of today's ride-on lawn mowers have an engine producing similar power ratings, and all they are required to achieve is a well groomed lawn!

One can talk about weight, torque, gearing, rear wheel

dimension and so forth. But the proof of the pudding would be to back-up a lawn mower to an F12, hitch them back-to back and see which one out pulled the other. Alternatively, try hitching a mouldboard plough to a lawnmower! There simply would be no competition.

A kerosene fuelled F12, subjected to various tests at the Nebraska Tractor Testing Authority in November 1933 (test no. 220) returned a drawbar pull of 1,814 pounds at 2.44 m.p.h. over a sustained period. It is doubtful if a lawnmower could have even moved the test rig!

An interesting innovative feature of the F12 was a pair of

*A tricycle Farmall F 12 restored by the author. Without mudguards (an optional extra at additional cost) the operating position was highly dangerous, owing the proximity of the lugs on the steel wheels.*

rods which were ensconced within the tractor chassis and connected the steering mechanism to the brakes. Whilst engaged in a full lock turn, perhaps at the end of a furrow, the turning of the steering wheel activated the relevant rear wheel brake, thus assisting with a tight turn.

There was however a down side to all of this. The two brakes were located inboard and were inadequate in the extreme for road use. Plus, the operator was obliged to lean forward to reach the two short hand leavers with two hands (there was no foot pedal) which meant releasing control of the steering wheel! Exciting stuff!

I used to own a tricycle three wheeled F12 and have vivid recollections of unloading it off the back of a truck and down a steep ramp!

Another less endearing feature of the International Farmall F12 was the fact that mudguards (fenders in U.S. parlance) were not offered as standard equipment. Accordingly a wise operator avoided wearing a loose fitting coat as the tall steel lugged wheels rotated worryingly close to the pan seating position. The wicked looking over hanging lugs extended beyond the wheel width and could easily drag a man to his death should a coat tail or sleeve be caught in their grasp – particularly if the variable wheel spacings were set to the narrow width setting! Also, an F12 operator certainly would *not want to nod off*!

The F12 was replaced in the latter half of 1938 by the F14. This was a near identical tractor but with the engine r.p.m increased to 1,650, giving the tractor a slightly improved performance.

An extensive array of implements were offered by International, specifically for attaching to each of the Farmall range. No less than fifteen cultivators were available, plus hay rakes, a quick attachable mower, numerous ploughs, to mention just a few.

In 1939 International released the first of the Farmall A series, a lightweight 19 belt h.p tractor. This proved to be the perfect unit for small acreage farmers. Numerous variations followed and indeed until the introduction of the Ferguson TE and TA series, the diminutive Farmall A became the world's top selling small tractor.

Also in 1939 the handsomely styled 27.9 belt h.p. Farmall H and the 36.7 belt h.p. Farmall M were introduced and, as with all Farmall models, became an instant success throughout the major farming countries of the word. The new row crop tractors surprised everyone by being decades ahead in styling, compared to the previous Regular, F12, F14 and the last of the original profile Farmalls, the F20.

The petrol/kero power plants installed in the H and M were undoubtedly the sweetest running tractor engines of the period. A diesel fuelled version of the M (the MD) became available in 1941 and although developing slightly less h.p. than the petrol/kero version, the extra torque of the engine increased the drawbar pull from 4,233 at 2.17 m.p.h. to 4,541 at 2.58 m.p.h.

Such was the popularity of the new big Farmalls, that, in order to ease the pressure at the parent Chicago plant, International's British implement factory at Doncaster commenced production of the H and M, identified by the letter 'B' which preceded the model type – example BM.

In 1951 International's Australian state-of-the-art new factory at Geelong also pushed ahead with the manufacturing of Farmall tractors. Their model types were preceded with the letter 'A'.

Driving an AM (in particular) was a joy. The big slow revving petrol/kero engine was almost vibrationless throughout its power curve. The well spaced 5 speed gear box was operated through a silky action clutch and a delight to use. An interesting feature was the well sprung seat, which could be tilted back enabling the operator to stand and stretch his legs, whilst continuing on with the job.

The Farmall saga extended for several decades with a bewildering range of model variations. Not all were available in every country. Regrettably, it would far exceed the confines of this epistle to detail the entire range.

But in view of the documented evidence of their past

outstanding integrity of design and their resulting reputation, it is not surprising that the present custodian of the name - Case IH - decided to once again introduce a catalogue of Farmall tractors to Australian farmers.

Bert Benjamin would have approved!

# POSTSCRIPT.

A necessary ingredient for an author who aspires to continue writing books is the receipt of feedback from readers and literary critics.  Fortunately in my case they are usually of a complimentary nature with only the occasional forthright acrimonious comment from an enthusiast choosing to disagree with my perhaps disparaging remarks relating to a specific tractor model, which happens to be his pride and joy! However I truly welcome all assessments, for these are my guideposts for future epistles.

I enjoy writing.  Words have always fascinated me.  I find it mentally stimulating to select from a myriad of descriptive adjectives and adverbs, the precise one which I believe will accurately convey my personal interpretation of a subject matter to my readers.

At the risk of sounding pietistic, I can truthfully state that my most gratifying reward I obtain from writing, is the knowledge that I am encouraging others around the world to take an interest in rescuing these grand old tractors from their places of  abandonment, for after all they are a part of a nation's heritage. Once unearthed, the prize invariably will be proudly carted home and treated to a caring process of restoration.

In addition, perhaps my narrative portrayal of distant places will encourage others to venture off the beaten track and explore intriguing but often overlooked corners of this amazing planet.

So, with my wife Margery at my side, I intend to continue heading off to distant horizons in pursuit of more travel and tractor tales, so that they may be shared with my readers, to whom I owe a considerable debt of gratitude.

*Ian M. Johnston.*